The Three Ethologies

A POSITIVE VISION FOR REBUILDING HUMAN-ANIMAL RELATIONSHIPS

Matthew Calarco

THE UNIVERSITY OF CHICAGO PRESS
CHICAGO & LONDON

The University of Chicago Press, Chicago 60637
The University of Chicago Press, Ltd., London

Published 2024
Printed in the United States of America

33 32 31 30 29 28 27 26 25 24   1 2 3 4 5

ISBN-13: 978-0-226-83243-2 (cloth)
ISBN-13: 978-0-226-83245-6 (paper)
ISBN-13: 978-0-226-83244-9 (e-book)
DOI: https://doi.org/10.7208/chicago/9780226832449.001.0001

Library of Congress Cataloging-in-Publication Data
Names: Calarco, Matthew, 1972- author.
Title: The three ethologies : a positive vision for rebuilding human-animal relationships / Matthew Calarco.
Other titles: Animal lives (University of Chicago. Press)
Description: Chicago : The University of Chicago Press, 2024. | Series: Animal lives | Includes bibliographical references and index.
Identifiers: LCCN 2023040935 | ISBN 9780226832432 (cloth) | ISBN 9780226832456 (paperback) | ISBN 9780226832449 (ebook)
Subjects: LCSH: Human-animal relationships.
Classification: LCC QL85 .C34135 2024 | DDC 590—dc23/eng/20231023
LC record available at https://lccn.loc.gov/2023040935

♾ This paper meets the requirements of ANSI/NISO Z39.48-1992 (Permanence of Paper).

Contents

INTRODUCTION

This book aims to propose a fresh vision for the future of animal studies. As a field of research and activism, animal studies has undergone tremendous growth over the past two decades. What was once a niche topic of interest to a few intellectuals and activists has found a generally well-established place in research paradigms across the humanities, social sciences, and physical sciences.[1] Further, the wide variety of approaches to animal studies has created opportunities for this research to come into dialogue with a number of leading movements for social and intellectual change, thereby expanding the field's focus and its intellectual and political concerns in ways vital for its maturity.[2] Such rapid expansion and development, though, has generated its own challenges. Amid the groundswell of animal studies publications, programs, certificates, and campaigns, it can be difficult to discern precisely what is at stake in this work and where the field might be headed.

One of the guiding threads that give coherence to much of the varied work in animal studies is the belief that dominant forms of human-animal interactions are a problem and need to be changed. In such practices as factory farming, invasive experimentation, and the use of animals in the entertainment industry and as companions (to name some of the more common modes of human-animal contact), human interactions with animals frequently involve unjustifiable exploitation, violence, and death. A significant part of the research and activism in animal studies is dedicated to reflecting on these harms and developing strategies to limit or abolish them. This generally critical disposition explains the field's heavy emphasis on restricting or prohibiting human interactions with animals (for instance, abstaining from consuming meat and dairy products, from using medicines and cosmetics tested on animals, from forms of entertainment that exploit animals, from buying animals from breeders, and so on) and the predominant interest in ethical and legal frameworks aimed at achieving equality and justice for animals.

In recent years, however, this primarily critical and restrictive approach to human-animal interactions has been supplemented by a more affirmative and experimental disposition that focuses not simply on restrictions but also on building alternative, more respectful, relationships with animals in a number of registers.[3] The continued development of this more experimental approach is crucial for the further maturation of animal studies, since prohibitions, restrictions, and abstentions (while entirely necessary in many cases) do not by themselves provide any positive vision for how else we might live. And surely one of the central aims of the field is to articulate genuine alternatives to the established order of human-animal relations. Moreover, focusing our critical perspectives on a handful of practices and institutional arrangements (such as factory farming, experimentation, hunting, and so on) has left unquestioned a whole host of other practices involving animals as well as the underlying values and ideals that structure our lives as a whole. A more affirmative approach to human-animal relations asks us to look at these unexamined dimensions of our lives and consider more fully how we might reform our lives as a whole. In other words, this approach raises the question of what a good life in common with animals might look like. And by articulating positive ideals of what forms good human-animal relations might take, we can also gain new perspectives on what modes of resistance are, in fact, worth pursuing and whether new modes might be necessary to achieve those ideals.

In the course of this book, I articulate certain key elements of this affirmative approach to human-animal relations by way of *three ethologies* that are aimed at transforming subjective, social, and environmental relations. This multipronged approach is inspired in part by Félix Guattari's brief but important work *The Three Ecologies*, a text written in view of the contemporary environmentalist movement.[4] Guattari argues in his book that the environmentalist movement should not limit its focus (as it has been prone to doing) strictly to contesting the degradation of the natural environment. Ecological problems can be adequately addressed, he maintains, only by developing a generalized ecological philosophy, or ecosophy, that is structured simultaneously in three ecological registers: mental, social, and environmental. These three registers correspond to what we might describe in more familiar terms as the domains of subjectivity, society, and the natural environment. Guattari's chief insight is that ecological issues are multidimensional in nature and hence require interconnected and transversal[5] responses that constitute a genuine alternative to the dominant culture. This framework (which has been refined and expanded in various ways by a number of other environmental theorists and activists) has served as a helpful corrective to environmentalisms that try to address ecological degradation primarily through hierarchical technocratic solutions as well as through reductive environmentalist frameworks that avoid engaging with broader social and psychological issues related to ecological issues.

In the chapters that follow, I propose that animal studies might profitably adopt a similar multipronged perspective in considering the stakes and the future of the field. In other words, I argue that contemporary pro-animal discourse and activism might be helpfully rearticulated as an *ethology* that has three forms: mental, social, and environmental. These three ethologies delimit and anticipate essential registers in which fundamental change is necessary; at the same time, they name and give form to work currently being carried out by activists and theorists in animal studies, especially those who seek to move beyond reductive frameworks focused primarily on legal reforms and consumer behavior and toward a broader vision of what human-animal relations might become.

The use of ethology within pro-animal discourse is, of course, not unusual; but engagements with ethology in this context have often been limited to employing its scientific findings to debunk outmoded pictures of animal behavior and cognition. Here I suggest a much broader understanding of what is involved in both doing and applying ethology, starting with a closer look at the term ethology itself. Typically, the term (from Greek *ēthos* + *logia*) refers to the study of animal behavior, a form of inquiry that is most often carried out from an evolutionary perspective and in the animal's natural environment rather than in a laboratory. There is, however, no reason to limit ourselves to this straightforward understanding of ethology as a chiefly scientific practice;

and, indeed, the term *ēthos* and the related concept of *ethos* originally carried a wide variety of additional meanings that are of interest to our project.[6] In ancient philosophical discourse, for example, *ethos* was understood to refer to the formed character or habituated dispositions of an individual; along these lines, we might consider the development of a *mental* ethology that investigates the various practices that form and reform the character and subjective constitution of an individual animal, whether human or more than human. The term *ēthos* typically signified custom, in the sense of the shared practices and relations that constitute a given social order; thus we could speak of a *social* ethology that aims to make a careful study and consideration of what constitutes social life among animals as well as between human beings and animals. In many ancient sources, *ēthos* also referred to the typical haunts or dwelling places of animals and human beings; thus we could speak of an *environmental* ethology that undertakes an analysis of human-animal relations within their environmental milieu. Here, then, we can see that the varied senses of the three ethologies forming the focus of this book are already contained within ancient uses of these terms.

The three ethologies we will be discussing are not, however, derived solely from etymological analysis but are also at work in ethological practice itself, especially among the practitioners of what is sometimes called "deep ethology." Let's first note some of the distinctives of traditional ethology and then work our way toward an understanding of this "deeper" version of ethology and its relation to our concerns. Ethology, as I noted above, is typically defined as the scientific study of animal behavior (1) as it occurs in its natural setting, and (2) in view of a given behavior's function and evolutionary history. The ethological study of animal behavior thus attempts to understand animal life in a more holistic manner than do competing schools like behaviorism. For scientists influenced by behaviorism and similar reductionist paradigms, animal behavior is ideally studied in the laboratory, often with little regard to a given animal's ecology or evolutionary history. For ethologists, by contrast, attending to the ecological, evolutionary, and relational aspects of an animal's behavior is essential to arriving at a full understanding of its significance.

Although ethology has obvious predecessors in the work of zoologists and naturalists from Aristotle to the Comte de Buffon to Charles Darwin, it emerges as a distinct field of academic inquiry only in the twentieth century, with the pioneering research of Konrad Lorenz and Nikolaas Tinbergen.[7] Tinbergen's "four questions" about animal behavior (which focus on behavioral causes, changes in behavior over a given animal's lifetime, reproductive fitness, and evolutionary history) are still frequently used to guide contemporary ethological research. The central idea animating this ethological framework is that multiple levels of scientific analysis should be integrated in order to arrive at a fuller understanding of a given animal's behavior, an approach that encourages practitioners to study animals from a situated and holistic perspective.[8] Building on the work of early ethologists, more recent practitioners have extended research on animal behavior into new registers. Gordon Burghardt, for example, has called for adding a fifth question to Tinbergen's four, with the additional question focusing on an animal's subjective, private experience (a topic we will explore at more length in chapter 1 under the rubric of mental ethology).[9] The aim of such work, which is sometimes referred to as cognitive ethology, is to understand animal cognition and subjective experience from the perspectives of animals themselves.

For cognitive ethologists of this sort, animals are best understood not simply as biological organisms to be studied using third-person, objective perspectives and methods but also as agents in their own right, who have their own centers of consciousness and who inhabit meaningful worlds.

It is not surprising that ethologists who study animal behavior with animals' subjective lives and perspectives firmly in mind often come to empathize with these animals and begin to see them as individuals who should be cared for and respected. Cognitive ethologist Marc Bekoff argues that the field of ethology should unabashedly affirm this affective disposition for animals and, in the process, should become a kind of "deep" ethology.[10] Bekoff uses the term deep ethology as an analogy with deep ecology, an approach to environmentalism arguing that ecology should involve more than a dispassionate scientific study of the natural environment and ultimately should base itself on a profound love of and respect for the natural world. Bekoff suggests that the same kind of ethical and emotive relation should ground ethology, thereby transforming the field into an investigation of animal behavior that is loving, caring, and compassionate—but still scientific and richly holistic. Deep ethology of this sort would, he maintains, help both scientists and nonscientists come to a fuller appreciation of the effects that human beings have on animals and make both groups more acutely aware of their ethical obligations to other animals.[11]

A number of notable ethologists have developed and adopted elements of this deep ethological approach. It is worth underscoring that many of the pioneers of this way of studying animals are women, and that their work has sometimes been dismissed based on the supposed sentimentalism of their methods (which often emphasize the subjectivity and relationality of animals in ways that mainstream scientific research tends to eschew). Some of the high-profile women scientists who clearly evince deep ethological sensibilities in their work include primatologists Jane Goodall, Dian Fossey, Biruté Galdikas, and Barbara Smuts (who has also done ethological work with other species), and elephant ethologists such as Cynthia Moss and Caitlin O'Connell.[12] In addition to these pioneering women, there are many dozens of researchers and activists who study animals using deep ethological sensibilities and whose findings have challenged scientists and the general public to expand their scientific and ethical appreciation of animal lives.[13] While taking a nontraditional approach to studying animals has led in some instances to these scientists' being excluded from more mainstream scientific circles, their work has also fundamentally changed long-standing ideas about animal life and has even led to the creation of new research paradigms.

In my home discipline of philosophy, there is a corresponding trend toward studying animals from a deeper, more generous ethological perspective. Under the names of philosophical ethology and critical ethology, a number of philosophers have begun to incorporate deep ethological practices and research into their philosophies and even to argue for a corresponding transformation of the field of philosophy itself.[14] Among the most prominent figures doing such work is French philosopher Dominique Lestel, whose substantial body of research is currently finding its way into English-speaking animal studies and philosophy.[15] Of particular relevance to this book are Lestel's efforts to combine ethology with ethnology and to narrow the gap between the human and animal sciences.[16] For Lestel, there is no sharp line dividing human and animal forms of sociality, and any genuine attempt to understand human sociality must of necessity incorporate its interspecific character.

Along similar lines, Belgian philosopher Vinciane Despret has developed a series of research projects highlighting the plasticity of human-animal sociality and what she terms "anthropo-zoo-genetic" practices.[17] Making extensive use of cutting-edge research in ethology, Despret argues that human-animal social relations can be and often are mutually transformative, thereby creating unprecedented forms of human and animal subjectivity and culture.[18] For philosophical ethologists of this sort, such experiments in human-animal sociality question long-standing anthropocentric prejudices in philosophy and the social sciences and call for the development of genuinely post-anthropocentric forms of thought and life.[19]

My aim here is not to provide a full summary of these intellectual and activist trends or to synthesize the vast body of research associated with deep ethology and philosophical ethology. Rather, I wish to underscore that the three ethologies developed in the main chapters of this book resonate with many of the central themes and concerns in deep and philosophical ethological research and practice. At a fundamental level, the arguments and framework developed here share with deep and philosophical ethologists the notion that human relations with animals are laden with normative significance and that a supposedly neutral or purely objective scientific approach to studying animals is neither desirable nor tenable in an age that is plagued by mass violence against animal life, biodiversity loss, and a rapidly deteriorating climate. One of the most important principles organizing this book and the ethological work that will occupy us here is that—for ethical, political, and ecological reasons—human-animal relations should be reoriented in better, more respectful directions. Thus the present project is animated by strongly normative concerns and is offered as a contribution to the research and activism associated with the aim of improving and reconstituting human-animal relations.

No less important than this normative orientation is the general ontological[20] approach to human-animal relations informing the bulk of this body of research and practice, an approach that starts from the premise that the respective natures of human beings and animals cannot be simply and easily distinguished. I refer to this approach using the term indistinction, a concept that is developed in view of the dissolution of classical versions of the human/animal distinction.[21] Traditionally, the distinction between human beings and animals has been characterized as sharply binary (that is, paradigm representatives of the human are assumed to have *X* ability or set of traits that animals lack) and normatively hierarchical (that is, having *X* is assumed to grant such human beings more value) in nature. The sort of ethological work I have just mentioned, supplemented by substantial research in animal philosophy and animal studies, has done much to undercut the assumption that human beings can be clearly and distinctly differentiated from animals in this way. Based on this work and on related intellectual and cultural trends, many people are coming to appreciate not simply that animals are "like us" in ways that entail equal normative consideration for them, but also that "we" (those individuals who consider themselves distinctively human) are like animals and are entangled in their lives and deaths in a variety of unanticipated ways that call into question traditional ideas about a distinctive human nature. The establishment and practice of the ethological approaches we will be examining here would be impossible without this sort of ontological sensibility and taste for indistinction. Indeed, the very notion that human-animal relations might be fundamentally transformed presupposes that there is some degree of plasticity in those relations and modes of existence, and that surprising and novel connections, assemblages, and bonds might be discovered and forged between and among human beings and animals that traditional conceptions of human and animal nature have often disavowed or denied a priori.

Overview

Each chapter in this book develops at some length one of the three ethologies while also considering linkages with the other two. Chapter 1 examines the notion of mental ethology by looking at the work of ethologist Joe Hutto. I examine Hutto's long-term experiment and sojourn with a herd of wild mule deer in Wyoming. Hutto lived with the herd for some seven years, seeking to understand its inner dynamics as well as the inner lives of individual deer. Hutto tells us that his relationship with the community of deer utterly transformed his sense of self, teaching him to displace his all too human perspective in favor of seeing the world "down the mountain" (from the perspective of the deer themselves). The subjective transformation Hutto undergoes is a striking example of mental ethology, a practice that I argue is crucial for challenging the anthropocentrism of the dominant culture. Of course Hutto's extreme ethological experiment cannot be easily replicated or universalized, but that is not at issue in this chapter. Rather, I examine his work to consider how sustained attention to the subjective lives of animals, and the concomitant transformation in our own sense of self that comes from such work, is an essential component in the constitution of better human-animal relations. The following two chapters take up other ethological experiments and encounters that proceed at different scales, but toward related ends.

In chapter 2, I discuss my relations with a murder of crows in my own neighborhood to explore the notion of social ethology. As I suggest in this chapter, the practice of social ethology embodies a vision of the social domain as something arrived at by experimenting with relationships rather than as a preexisting, static sphere to be discovered. This more open-ended approach to sociality challenges the anthropocentrism of traditional ideas about culture as restricted to interhuman relations. At the same time, this view raises the question of how far such reconfigurations of sociality might extend and the possible consequences such expansions might have for urban life. In view of these issues, I engage with a variety of recent theorists who have developed and refined the "political" turn in animal studies. Through such notions as "zoopolis" and "companion species," these figures provide us with important conceptual and practical tools for thinking about how urban life might be transformed by a more generous vision of human-animal sociality.

Chapter 3 focuses on environmental ethology, where the concept of environment comprises both the ecology and the land. I use "land" here in the sense used by resurgent Indigenous writers, who understand it to refer to an extensive set of relations that binds the human and more than human worlds. As a means of thinking through these two registers together in relation to animal life, I turn to Indigenous writer Linda Hogan (Chickasaw), whose extensive body of writing has long been situated at this complex intersection. Taking her recent work on animals as a point of departure, I then examine two particularly pressing ecological issues affecting animals today—mass extinction and climate change—in order to think about these issues from a non-anthropocentric perspective. Moving beyond ecological issues, I inquire how the notion of land as developed by resurgent Indigenous thinkers and activists might be brought to bear on contemporary pro-animal discourse and activism. I suggest that this notion of land raises two issues that are fundamental to contemporary animal studies: the scope of relations with the more than human world and the way pro-animal discourse and activism are implicated in settler colonialism. The chapter closes with a brief analysis of Mi'kmaq scholar Margaret Robinson's approach to veganism, in which she offers a helpful example of how to think more carefully about the complex set of tensions and affiliations between pro-animal and Indigenous ideas about animals and land.

The conclusion explores how these three ethologies are animated by a practical ideal that I refer to as *syn-theōria*, or seeing-with animals. This ideal, I suggest, forms the core of a philosophical way of life that aims to recombine and reconstitute traditional notions of the good, the true, and the beautiful. I conclude by suggesting that a worthwhile life might be found today if we risk opening ourselves to the admixture of transformative joy and profound grief that characterizes contemporary life and death, in common with our animal kin.

A Note to Readers

I have tried to make this book's main arguments and concepts accessible to a broad range of readers who are interested in animal studies but are not necessarily specialists in the field's contemporary philosophical and theoretical debates. As with any work of philosophy, you will find arguments that take some heavy lifting. I am confident, though, that if you persist, you will understand the main ideas. More important,

I hope you will find them useful for rethinking interactions with our fellow animals.

For scholarly and specialist readers in animal studies and related fields, I hope this book—although written with non-specialists in mind—will also be of use in your own research and practices. The ideas here are the fruit of many years of reflection and practice, and I believe they offer useful ways of reframing and refreshing central issues in animal studies. If particular arguments or ideas are not fleshed out here to the extent that would be standard in a scholarly monograph, or if certain materials from the primary or secondary literature have gone missing, I hope you will consider consulting the more traditional scholarship I have published elsewhere.

CHAPTER ONE

Mental Ethology

The Inner Lives of Animals

"I have lived with a large herd of mule deer . . . every day for the past seven years. How, you may ask, can a person do such a thing, and why would he choose to do it? My response would be, how could you not? Given the opportunity, how could a person resist such a life?"[1] Thus begins field biologist Joe Hutto's account of one of the most remarkable deep ethological experiments in recent times. Hutto's experiment took place in Wyoming, where he lived with his wife and research partner, Leslye, on an old homestead in the foothills outside the town of Lander. Hutto had undertaken a previous experiment with bighorn sheep in Wyoming and decided to return to the area for further research.[2] He had no idea that the homestead was within the winter range of a herd of mule deer. On most days Hutto's experiment took him high into the surrounding mountains to study the daily lives and rhythms of the herd. During this extended period of living with the deer, Hutto developed surprising insights about these animals and also underwent a profound transformation in his own sense of self.

Hutto didn't decide to engage in a deep ethological study of the deer before coming to the homestead. It was chance interactions with the deer themselves that prompted him to think such an experiment might be possible. Out walking in the sagebrush one day, Hutto encountered a buck who was staring at him with apparent interest. He carefully glanced back at the buck, spoke a few words, and made a gesture with his head. The buck didn't run but instead returned the gesture with his own head. This encounter made Hutto wonder what other kinds of interaction with the deer might be possible if he were to place himself in their midst.

Later, while standing in their homestead kitchen, Hutto and Leslye noticed a nearby doe watching them. They returned her gaze and, much to their surprise, she didn't flee; instead, she continued to walk casually around the ranch, going about her business. Rayme (the name Hutto and Leslye gave the doe) began to visit the ranch each afternoon, bringing her family with her. Hutto suspected that Rayme had somehow been close to other humans before visiting the ranch, but none of the surrounding ranchers had noticed this. Rayme continued to show a pointed interest in Joe and Leslye and would stare through their open windows for long periods. Hutto explains that "Rayme somehow managed to make it perfectly clear not only that she was consumed with interest regarding our activities but also that she clearly wanted attention from us. There was something about our mere proximity that Rayme desired. For reasons that will always remain a mystery, Rayme found us—Rayme sought us out" (28).

These and other such encounters with the deer gave Hutto the idea of seeking closer contact with Rayme's family and the herd as a whole. He knew, though, that such intimate interactions would not occur overnight. It took about two years of being near the herd every day for them to accept that Hutto was not a threat. Moreover, he was not fully adopted by the herd until one doe in particular approached him and made contact. Hutto named her Raggedy Ann because she had experienced so much strife in her life. He describes her as being extremely nervous and always maintaining a cautious distance from him. But one day, after Hutto had humbly followed the herd at a distance, Raggedy Ann approached him. He describes this as a "profound" and "amazing" moment in which Raggedy Ann's body relaxed, her eyes softened, and she walked straight up and touched him.[3] She had effectively now decided Hutto was safe to have around. What Hutto did not anticipate was that Raggedy Ann's gesture would transform the whole herd. Once she accepted him, most of the other deer did so as well. Hutto gradually came to understand that Raggedy Ann was the matriarch and that she was regularly making important decisions for the whole group, including deciding to accept Hutto as a full member.

Throughout his account of the experiment, Hutto is keen to stress that accepting him was deeply atypical behavior for the deer. He notes that even when he is standing in the midst of the herd, the deer remain extremely nervous about the presence of other human beings and will bolt at the slightest provocation. The deer thus recognize Hutto as a unique individual in a unique relationship with the herd; they recognize his particular body, movements, and voice and respond differently to him than to all other human beings. Conversely, Hutto learns over time to see each deer in the herd (some two hundred) as a distinct individual, with its own body markings, disposition, and personality. This respect and attentiveness to the singularity of the deer helps him to establish different relations with each individual. Some of the deer show little interest in Hutto and simply tolerate his presence, while others have become close friends who seek out his presence and attention. Hutto's book and documentary are filled with extraordinarily moving, loving, and heartbreaking accounts of these relationships. Because of his proximity to the herd, Hutto experienced everything from a doe's inviting him to witness the birth of her fawns to being alongside the deer as they were hunted and attacked by animals and human beings alike.

These close relationships and sustained interactions with the herd made it impossible for Hutto to remain an objective scientist during his experiment. He is forthright in acknowledging that his work crosses scientific lines and travels into philosophical, poetic, and artistic registers. His account regularly mixes fine-grained objective, scientific analysis with subjective, qualitative impressions; he also offers frequent reflections on his own emotional state during the experiment, which ranges from utter bliss to deep despair. Hutto risks taking a nontraditional approach to studying the deer because his ethological experiments are driven by ends rather different from those of academic ethology. Hutto is not just interested in studying animal behavior in natural habitats; nor is he content simply to grant that many animals have consciousness and their own perspective. He aims to explore the inner lives of animals and to get as close as possible to seeing the world from an animal's perspective.

Down the Mountain

Hutto, of course, is not alone in studying animals in a way that challenges the boundaries of traditional animal science.

He traces his unorthodox ethological approach to several predecessors I mentioned in the introduction, including the primatologists Jane Goodall and Dian Fossey, as well as one of the founders of modern ethology, Konrad Lorenz (284). What renders Hutto's work particularly worthy of our attention here is his special emphasis on trying to gain access to the inner lives of animals as well as the changes he undergoes in undertaking such studies. In terms of animals having their own subjectivities, Hutto's extended embedding in the herd convinced him not only that the deer have the bodily and cognitive components of rudimentary consciousness (what he calls the "provisions of consciousness" [283]), but that they are aware of and awake to the present in ways that match and even exceed the experience of most human beings. As Hutto emphasizes, the deer's very survival depends on their being acutely sensitive to innumerable factors in their environment, from the changing seasons to variously distributed food sources to the different personalities and behaviors of other members of their herd. Further, Hutto has no qualms about attributing to the deer a heightened awareness of death. As a prey species and as regular victims of human hunting and road accidents, the deer are more familiar with death than many human beings are. They regularly witness herd members being attacked and killed, and in their daily travels they frequently pass the remains of dead deer.[4] Hutto describes at length the painful grieving that a dead deer's family members undergo when one of their number dies; he also details the ripple effects such deaths can have on those family members and even on the herd as a whole.

Beyond the task of describing animals' inner lives based on external evidence, Hutto's goal is ultimately to enter into those lives and get a feel for the lived experience of the deer themselves. By embedding himself in the herd and developing close relations with several deer, Hutto becomes closely attuned to the rhythm of daily life among the herd. He and the deer learn to share the daily joy and suffering of life and death. In the process, Hutto is graced with a very different perspective on the world. He learns, as he says, to see "down the mountain" (from the deer's perspective), and not just "up the mountain" (from a typical human perspective).

In working his way into the inner lives of the deer, Hutto undergoes a parallel conversion in his own sense of self. It is important to recall that on most days Hutto spent fourteen hours with the herd. In so doing, he experienced a literal and figurative deterritorialization (to borrow a term from Gilles Deleuze and Félix Guattari).[5] Thus, unlike most field biologists, he was not studying the deer using cameras and binoculars from a comfortable distance in a research station. Instead, he met the deer on their terrain, in their home, and walked countless miles with them in the hills and mountains. Wild deer are extremely skittish, so Hutto had to learn to carry his body differently. Arrogance or swagger was out of the question. He learned to be gentler, humbler, more patient, and more deliberate in his movements; he studied how the deer interacted with each other and took cues on proper behavior from their gestures and vocal communication. Hutto had to become-deer, then, in a manner pushing up to and perhaps beyond the limit of what Deleuze and Guattari call becoming-animal.[6] For Hutto's experiment to accomplish its aims, he had to do more than inhabit a realm of shared affect (the focus of Deleuze and Guattari's analysis of becoming-animal); he had to adopt the herd's very form of life.[7] Over time, this ethological work sank into his subjective constitution at a habitual and bodily level, redirecting his entire way of seeing, being in, and moving through the world.

At the outset of his experiment, Hutto saw himself primarily as a rancher, a land manager, and a man of science. He was also an avowed hunter and meat eater and had spent many years killing and eating deer with little consideration of their experience of being attacked and hunted. Over the course of the seven-year experiment, however, Hutto experienced significant changes in both his point of view on the world and his emotional constitution. He became less and less able to study the deer in an objective manner or to take a detached perspective on their lives and deaths. Instead, he came to view the deer as his kin, loving them much the way people love their family members and close friends. Thus, when one of the members of the herd dies, that death is for Hutto not just *akin to* the death of a family member—it *is* the death of a family member. And when the herd is in the crosshairs of a gunsight, Hutto sees it not from the perspective of the human hunter but from the perspective of the deer who are under threat.

Hutto's relations with the deer thus brought him into direct conflict with the established subjectivity he carried with him into the experiment. Throughout his written account, for example, he insists on the legitimacy of hunting mule deer, wishing to maintain this practice in a sustainable manner. Yet as he grows increasingly attached to the deer both as individuals and as a herd, his predilection for hunting is undercut from within. He explains that he now finds himself faced with an irreconcilable paradox, unable to view hunting primarily from the human perspective he previously occupied. When he sees one of the mule deer killed by a hunter, he sees and feels that death along with the deer, immersed in the herd's suffering and grief:

> Now, when a bullet passes through the body of one of my family members . . . there is no more displacement or refuge from my attachment—that mindless, objective space where previously my emotions would have safely resided. Now there is only the shared pain and agony and the loss of one that I care for deeply. (294)

In being transformed by sharing the daily life of the mule deer, Hutto becomes dehabituated from his usual ways of seeing and reoriented into a new economy of subjectivity. His established identities slowly give way to new conceptions of self and relations, and his daily interactions with the deer create alternative forms of individuation. Toward the end of his reflections, Hutto writes movingly of how transformative his ethological project has been and how thoroughly his subjective constitution has been reoriented in the process:

> I am reminded that my life has been thoroughly interwoven with this phenomenal animal for so many years. And after sharing so much history, it is hard to know how this has changed me; it is hard to remember who I once was, and harder still to understand who or what I may have become. My identity has undeniably been reshaped and redefined by this community—this family into which I have in some strange way been assimilated. Increasingly, it appears that my world—my frame of reference—has been irrevocably changed. Perhaps, at last, I am in fact seeing a different perspective—seeing the world through another creature's eyes. (294)

Hutto's relationship with the deer has far-reaching effects for the larger stakes of his project. An experiment that began out of scientific curiosity and a certain affection for the deer becomes, by its conclusion, an impassioned plea to his fellow human beings to save the deer from the ravages of modern human society and ecological degradation. Once the perspective of seeing "down the mountain" becomes a regular practice, Hutto is compelled to extend his reflections beyond the boundaries of scientific analysis into the social and environmental registers that will occupy us in subsequent chapters. He comes to appreciate, at both a cognitive and a visceral level, that the deer will not survive if the dominant culture continues on its current path. Thus Hutto's work is offered to his audience as an urgent call to change our social structures and to adopt saner and more sustainable ecological habits. He shares his subjective conversion with us as a witness to another perspective on, and another way of life with, the mule deer that makes their sustained existence possible in the face of the ongoing threat of their extinction.

What Is Mental Ethology?

With this brief overview of Hutto's experiment in mind, we can now schematize what is at stake in practicing mental ethology. Let's begin by defining mental ethology with a bit more precision and then flesh out some of its implications. In the most general sense, mental ethology refers to the study of subjectivity among animals. The focus here is on what Hutto labels "the inner lives of animals"; thus, mental ethology proceeds on the assumption that animals have their own subjective points of view and that to some extent it is possible to gain access to these perspectives. Mental ethology is also a matter of considering the conditions under which processes of subjective formation are either enhanced or degraded. For example, how might fragmented habitats and ecological degradation limit the scope and range of possible relations a given animal might encounter? Conversely, how might improved living conditions allow for a broader range of subjective potentials for a given animal?

Implicit in this initial definition of mental ethology is the notion that animal subjectivities of the sort under consideration are *singular* and *plastic*. Singular here means a given animal's subjective constitution will vary based on the specific relations and conditions for subjectifying in which it finds itself. Plastic means that an animal's subjectivity carries a certain flexibility. Subjectivity does not come ready-made. It is forged through various experiences, encounters, and relations. I should emphasize that the attention to the uniqueness and contingency of subjectivity is by no means premised on ignoring the phylogenetic heritage that structures an animal's existence. Hutto repeatedly emphasizes that the deer he interacts with carry a powerful set of instinctual dispositions that constrain their behavior in fundamental ways. Deep patterns and structures persist among the herd, even in the face of powerful counter-influences from the human and more than human worlds. The same is true, of course, of other animals we might interact with (humans included).

One of the paradoxical effects of studying animals from a mental ethological perspective is that it transforms both the human and the more than human actors. To practice mental ethology in the sense exemplified by Hutto and his deep ethological predecessors is to abandon the safe confines of the research laboratory and the relative distance of the field station and to become one of the points of contact through which animal subjectivities are formed. Mental ethology thus involves the researcher's going beyond the abstract study of animals' inner lives to become an active participant in forming those inner lives. This formative process, as I noted, runs in both directions. Entering into intersubjective relations with other animals entails that we, too, undergo subjective changes. The sorts of experiences, encounters, and relations that constitute our everyday lives will change, and our subjectivities will be exposed to the potentials and possibilities inherent in those shared spaces.

Through the daily routines of being-with and seeing-with other animals, the practice of mental ethology gradually gives form to another sense of self. If most of us are trained from birth to orient ourselves almost exclusively toward interhuman affairs and relations, mental ethology serves to orient us in new directions and to train our attention on affairs that are more than human. But for this sort of reorientation to take genuine hold on our subjective constitution is no mean feat. In Hutto's case, it took years of building an alternative set of daily habits and routines alongside the herd for the practice of seeing-with the deer to become a settled disposition. Such a commitment to reorienting oneself might seem extreme and overly burdensome; but if we reflect on the long-term processes by which we are formed to be all too human subjects, it should not surprise us that becoming a different sort of person requires enacting countermeasures that are similarly sustained and prolonged.

Hutto's reorientation toward affairs that are more than human ultimately leads him to develop a profound love for the deer and to have an existential stake in their lives and deaths. Being-with and seeing-with the deer every day makes it impossible for him to maintain the subjective posture with which he began the experiment. Hutto leaves the experiment utterly transformed. His critical reflections on his own eating patterns and former hunting practices, along with his urgent and heartfelt pleas for collective action to save the mule deer from extinction, thus do not stem from an abstract intellectual commitment to animal rights or animal welfare. They grow out of his love for the deer and his existential stake in their continued existence and flourishing.

You Must Change Your Life

Let's pause and take stock of the path we have traveled thus far. I have explored Hutto's deep ethological engagement with wild mule deer primarily in view of its implications for what I have called mental ethology. The aim of his experiment is to gain a glimpse of the inner lives of animals; and to do so, he must learn to occupy the deer's point of view as much as he can. In entering this shared space and gaining an alternative perspective on the world, Hutto finds himself changed. A man who begins with the perspective of a rancher, scientist, and manager and who initially looks "up the mountain" learns to see "down the mountain," from the perspective of the herd. The daily practice of seeing and living with and from the herd's standpoint ultimately leads to a shift in his affective and relational commitments. Hutto can no longer remain aloof and stand by passively as the deer are subjected to suffering, violence, and death. At this point, as he searches for ways to respond to their needless suffering, Hutto's practice of mental ethology finds itself opening onto the social and environmental ethologies we will explore in subsequent chapters.

Before charting those connections, however, our preliminary task is to examine in more detail how the practice of mental ethology might be relevant to contemporary work in animal philosophy and animal studies. We can better appreciate the alternative trajectory opened up by a mental ethological approach by first contrasting it with the general direction and aims of much of the work carried out in the context of contemporary pro-animal discourse and activism.

The dominant paradigm within pro-animal discourse and activism is premised on the idea that the central reason humans mistreat animals is *speciesism*. This term, coined by Richard Ryder and popularized by Peter Singer and others, refers to a widespread and supposedly "irrational" prejudice that leads human individuals to limit ethical consideration to members of their own species.[8] The chief limitation with speciesism, these thinkers suggest, is that when we carefully reflect on the criteria we typically employ to include our fellow human beings within the scope of consideration (criteria that range from subjectivity to sentience to autonomy), we see straightaway that these capacities and characteristics are not exclusive to human beings but can be found in varying degrees among some animal species. Pro-animal critics further assume that once this speciesist inconsistency is brought to light, most of us will be moved (on pain of logical inconsistency) to reconsider the many unjustifiable practices grounded in this prejudice. On their account, this reconsideration should ideally issue in a combination of ethical reforms at the level of the individual (for example, switching to a vegetarian diet, not buying products tested on animals, and so on) and broader collective efforts aimed at legal and policy changes (for example, laws that support and expand animal welfare or rights) that encode non-speciesist and logically cogent norms in policy and law.

There is much to be said in support of this general approach to pro-animal discourse and practice. The critique of speciesism is focused on discerning the ultimate source of violence against animals while also being concerned to limit and even abolish the most egregious forms of such violence from our individual and collective lives. These are important critical and ethico-political goals, and this book is indebted to the several decades of research and activism carried out in this vein. That said, this standard approach is plagued by several problems, a few of which I will emphasize in the present context.[9]

Consider, first, the concept of speciesism itself. Recall that speciesism is defined as an irrational prejudice that leads us to limit ethical consideration strictly to members of our own species, *Homo sapiens*. Although one can find instances where certain individuals hold the position that all members of the human species have moral standing to the exclusion of all nonhuman animal species, dominant normative attitudes and practices concerning who does and does not matter are typically far less clear-cut than this. Persistent and prevalent forms of sexism, racism, colonialism, ableism, and countless other forms of marginalization give the lie to the notion that the human species as a whole is generally included within the normative scope of the established ethical and social orders; moreover, even the most robust and expansive human rights discourses and laws are plagued by intractable problems concerning what constitutes humanity proper (for example, in the context of beginning- and end-of-life debates), raising the question whether even human rights frameworks are truly inclusive. Conversely, even though countless animals are subjected to horrific violence, suffering, and death, some animals *do* have ethical, political, and legal standing and protections in a variety of personal and institutional contexts. In short, neither the dominant culture nor the dominant ideology concerning animals can be accurately described as speciesist.

A more accurate description of the hegemonic social and ideological order would be that it is *anthropocentric*.

Although this term is clunky, it is quite useful, as it nicely captures a persistent pattern that has characterized the established social order (especially in what we might call "the West") for millennia.[10] So, precisely what does it mean to characterize something or someone as anthropocentric? Literally, the term means that the human (Greek *ho anthrōpos*) occupies the center; thus, to be anthropocentric is to be human-centered. (This definition might initially look much the same as the definition of speciesism, but the differences between the two concepts will emerge in what follows.) Such human-centeredness can be found in a wide variety of common attitudes, recurrent practices, and influential institutions. In regard to attitudes toward animals, anthropocentrism typically involves drawing a distinction between "the human" and "the animal" as well as "the nonhuman," maintaining that full human beings have special capacities and characteristics that animals and other nonhuman beings lack (ranging from consciousness and language to moral agency and politics). The lack of such capacities and characteristics among animals and other nonhuman beings is often (but not always) assumed to give them lesser value than human beings and sometimes no worth at all.

In contrast to the standard characterization of speciesism as constituting a set of beliefs or attitudes, it is important to recognize that anthropocentrism encompasses much more than a mere set of intellectual commitments. It is an entire way of being in the world that is instituted and sustained through innumerable practices cutting across every register of social life—from what we eat and the medicines we use to how we move and inhabit the places where we live. The division between human and animal, and between human and nonhuman, is enacted and given material reality in and through such practices. One's thinking and subjective disposition thus become anthropocentric by one's being formed and raised within these practical and institutional milieus, unconsciously absorbing the common habits and practices that circulate in and through them. By the time we might become reflectively aware of anthropocentrism and construct explicit beliefs about it, we have already been formed for many years (indeed, since birth) as anthropocentric subjects. Undoing such anthropocentric subjectification, then, cannot be limited simply to making changes in our conscious beliefs or attitudes; it requires, more fundamentally, adopting another mode of subjectivity and another form of life.

There are other crucial differences between anthropocentrism and speciesism. First, the notion of "species" at work in the term speciesism refers primarily to a biological concept of species that is fairly recent in origin. The problem with speciesism for pro-animal thinkers, as we saw, is that it invokes membership in the biological species *Homo sapiens* as the chief criterion for ethical consideration; these thinkers go on to argue that such biological markers are ultimately irrelevant when determining who should receive ethical consideration (the assumption being that other markers such as sentience or autonomy that cut across biological species lines are more relevant). But the marginalization and exclusion of animals and nonhuman beings from the dominant ethical and political community can be found long before modern notions of species were established, going back to the earliest surviving documents and archaeological evidence in Western history.[11] People from these earlier periods had no concept of the modern notion of biological species and understood humanity in a variety of nonbiological ways that allowed for a division between the human and its animal and nonhuman others. And even if many people today are generally familiar with biological ideas about species, their practices and ideologies are not typically founded on rigorously biological divisions between *Homo sapiens* and other species. Far more common are the divisions drawn between what is taken to be *fully and properly human*, on the one hand, and those beings and things considered to be *nonhuman* and *subhuman*, on the other. (The latter categories often include an admixture of human, animal, and other nonhuman beings.) In this sense, "the human" tends to serve less as a biological species marker and more as a kind of *subject position*—that is, a normative ideal of what constitutes a full and proper human being—to which only a select group of human individuals ever accede and from which vast parts of humanity (and most of the nonhuman world) are structurally barred.

In brief, then, what is primarily at stake in distinguishing the human from the animal and the nonhuman, both historically and in our present age, is not so much clarifying the boundaries of biological species membership as determining who belongs within the scope of *anthrōpos*, or the human proper—those individuals and groups who are taken to be truly, fully, and genuinely human. This is why Western culture has never in the past or in the present had a problem with subdividing human beings from within. Our history and our present era are replete with examples of the fracturing of humanity (dehumanization, subhumanization, animalization, and so on) in which individuals who are members of what biologists would label the species *Homo sapiens* are portrayed as not fully and properly *human*. We are entirely familiar with the discriminatory gesture of using the term human to refer to a supposedly universal humanity while simultaneously denying that very humanity to slaves, women, children, colonized peoples, and certain ethnic and racial groups. The concept of anthropocentrism helps us to see that one of the persistent patterns of Western culture is a simultaneous division of the human from the animal and an internal division and fracturing of the human that places dehumanized and subhumanized groups in a liminal space in relation to the orbit of the human proper.[12]

A concomitant problem with the standard approach to speciesism, mentioned briefly above, is that it encourages us to see the mistreatment of animals as primarily the consequence of an intellectual failure. Speciesism is often portrayed (especially by scholars who place a premium on logic and philosophical argument) as an irrational belief plagued by logical inconsistencies; and the way to correct such muddled thinking and its correlative practices, the standard approach goes, is by careful, critical thinking. Now, it would be pointless to deny that there is an intellectual component at work in upholding dominant ideologies and practices regarding animals, and that our ideas in this domain are often muddled and inconsistent. Likewise, it is surely true that careful and critical reflection on our ill-founded and dogmatic beliefs can motivate us to develop better ideas and better-grounded practices. But if we bear in mind the notion of anthropocentrism just developed, it should be clear that we did not arrive at dominant ideologies and practices regarding animals primarily by way of irrational and muddled thinking; indeed, reflective thinking has played little role in the matter at all.

Here, as in many other issues related to subjective formation, scholars (and especially philosophers) have been beguiled by what we might call an *intellectualist* paradigm that sees the preponderance of human behavior as driven by reflective intellectual commitments. Although this picture of human motivation might be true of a small class of our behaviors, it is certainly not true for the vast majority of our interactions with animals and our attitudes toward them. Most of us grow up within an anthropocentric form of life and learn to take on an embodied set of habits and daily practices that train us from birth to perceive anthropocentrism as natural and inevitable. To be sure, if this web of practices and beliefs is laid out as a series of propositions and intellectual commitments, it is manifestly absurd and unjust in countless ways. But intellectual recognition of the logical and ethical problems with anthropocentrism will not, by itself, constitute enough of a force to change our subjective constitution. Our subjectivity has been built, in some cases for decades, through the adoption of subtle anthropocentric habits, practices, and ideologies that penetrate every facet of our lives. To unravel this sort of subjective formation and to become the kinds of subjects who militate for justice and a better way of life, and who remain faithful to such a project for a lifetime, requires us to undergo sustained subjective countertraining.

Excursus: Soulcraft in Foucault and the Cynics

One philosopher who is particularly helpful for illuminating both aspects of this countertraining—both the initial constitution of subjectivity by the established social order and the processes involved in trying to contest and reshape that subjective formation—is Michel Foucault. Foucault is justly famous for his analysis of the first aspect of subjectivity, especially in terms of the role dominant social modes of power play in subjective formation. A substantial portion of his most widely read writings focus on challenging the commonsense notion that human subjectivity somehow preexists or escapes the influence of intersubjective and social relations.[13] As Foucault repeatedly and persuasively demonstrates in these writings, human subjects are formed from the ground up through a variety of daily practices and institutional mechanisms of power—apparatuses and relations that he explores under such rubrics as discipline, governmentality, and biopolitics.

The chief lesson of Foucault's analysis of subjective formation is that there is no possibility of uncovering some authentic or genuine self by digging back behind or outside power relations. To believe we might be able to recover a true self in this escapist manner is to be misled by what Foucault calls a "confessional" notion of the self—a notion that encourages us to think we can somehow "confess" or express a deep inner self free from the contamination of discipline, external forces, or influences. But if we are in fact subjects of power—that is, if we are selves built from the ground up in and through power relations of various sorts—then all that introspection will ultimately disclose are those power-laden relations. Our true, authentic selves simply *are* those subjects of power, not some fictive selves behind or outside the orbit of power-laden relations. As should be plain, anthropocentrism as I have analyzed it in this chapter functions very much like the power relations Foucault highlights. Anthropocentrism forms us as subjects from the ground up, and there is no way for us to find a true self outside or behind its influence.

In the final years of his life, Foucault appeared to change tack, approaching subjectivity from another angle. Rather than merely exploring the ways the established order forms us as disciplinary subjects, he became interested in the re-formation of subjectivity, or the intentional practice of *care of the self*, a concept he developed through a sustained engagement with ancient Greek philosophy (among other sources).[14] Foucault is sometimes criticized for being inconsistent in making this turn to a more affirmative notion of the self, an accusation that assumes the bulk of his previous work was spent trying to utterly dissolve selfhood and subjectivity into social relations and structures. But there is nothing inconsistent about Foucault's interest in practices of care of the self, for such care does not amount to care for an authentic self that escapes history, power, or social formation. Rather, care of the self is the patient practice of attending to the conditions under which a given subjective formation might be re-formed or counterformed given the materials and relations at hand. Thus, care of the self—or what we could call *resubjectification*—takes as its method and goal something very similar to what I have described in analyzing mental ethology.

For Foucault, the ultimate aim of resubjectification is getting "free of oneself" and becoming a subject different from the one formed by dominant modes of subjectification.[15] The strategy is thus to figure out what concrete practices and habits we might undertake in order to shape our subjectivities in a new manner. Following the groundbreaking work of ancient philosophy scholar Pierre Hadot, Foucault sometimes refers to these practices as *spiritual exercises*, insofar as they are aimed at a spiritual (or subjective) conversion and reconstitution.[16] Resubjectification does not, then, take place "from scratch," as if we were working with a clean slate imprinted with whatever new form we might desire. Rather, resubjectification is a matter of converting and transforming the subjective constitution we have at hand as well as the relations and institutions within which we have been formed. There are two relevant questions from this perspective: What sorts of novel subjective formations do extant subjective and social conditions render possible? How might existing practices and techniques of subjectification be repurposed and reformed to alternative ends?

The bulk of Foucault's writing on these themes tends to characterize the processes of resubjectification in exclusively interhuman terms: he typically locates power in interhuman relations and institutions and looks for sources of resubjectification in those same registers. He rarely considers the possibility we are exploring in this chapter, that the modes and sources of resubjectification might have a considerably wider scope that extends to the more than human world. Just before his death, however, in his final lecture course at the Collège de France Foucault takes up a sustained analysis of the ancient Cynics in which he glimpses something of this broader vision of resubjectification.[17] The Cynics were radical countercultural philosophers for whom animals and the more than human world stood as exemplars of the good life. They argued that our subjective lives tend to be largely indexed to the limited ends, ideals, and relations promoted by the dominant human culture. They insist, though, that interhuman and cultural modes of power and relation are not total; we are also formed by and exposed to other sorts of subjective potentials in the animal and the more than human worlds. The Cynics, then, do not restrict themselves to strategies of resubjectification based on reworking and repurposing dominant cultural norms; instead they expand the pool of potentials for transformation by exploring the variety of possibilities for subjective formation among animal life and the rest of the more than human world. It is this gesture of seeking possibilities for resubjectification among the more than human world that is distinctively Cynical, a strategy that intrigued Foucault in the final portion of his life.

The name the ancient Cynics gave to this expanded approach to resubjectification is "living according to nature" (*kata physin*). To live according to nature in the Cynics' sense is to contest the primacy the established order grants to human cultural norms (*nomoi*) and to give a higher and "positive value to animality."[18] As Foucault notes, indexing one's form of life to animality and animals is a scandalous gesture in the ancient context, for the sort of anthropocentrism identified in this chapter is already at work there. In the ancient Greek context, animals and animality are commonly understood to belong to a zone outside humanity proper, with animal existence marking a "point of repulsion" in the constitution of the human.[19] And for most ancient Greek philosophical schools, human flourishing is taken to depend on human beings' discerning their unique mode of existence as opposed to animals and maximizing that distinguishing characteristic, whether it be human rationality or human ethico-political norms.

The Cynics, by contrast, maintain that this kind of human exceptionalism and anthropocentric navel-gazing is precisely the reason human beings fail to flourish. The second natures we have acquired through dominant cultural modes of subjective formation do not provide us with lasting happiness or with the character traits for living a worthwhile life. Animals, they argue, tend to be far happier than we are; they are content with very little and have become so accustomed to the vicissitudes of life that they are able to endure and flourish under natural conditions.[20] This is why, as Foucault notes, the Cynics take animals as both sources and exemplars for the task of resubjectification. The Cynics pay close attention to how animals live and strive in order to take on the character traits and virtues that animals exhibit. Achieving this alternative subjectivity thus involves deliberately forming a *third* nature—a new ethos and character—that is brought about by rehabituation and repeated practice (*askēsis*); it requires becoming dehabituated from the second natures we have acquired through culture and civilized life and patiently building a third nature in relation to and in view of animals and other more than human entities. This third nature is intended to re-form the self and is also held out to one's fellow human beings as a scandalous challenge to find genuine happiness in life by becoming more like our animal kin. It is in this sense that Foucault can describe animality as an exercise and an achievement for the Cynics. Animality of this sort does not come naturally to human beings, nor is it a state we can revert to; it is the result of intentionally adopting another way of life, in view of the limited scope of possibilities for subjective formation offered by the established order.

In emphasizing the positive prospects and potentials for resubjectification with and among animals and the more than human world in the ancient Cynics, Foucault points us toward an early example of the practice of mental ethology. Although he does not provide a complete genealogy of this expanded vision of resubjectification, we should note that numerous forms of life throughout the history of both Western and non-Western philosophical traditions prefigure mental ethology by indexing flourishing to a life lived according to animals and nature.[21] This minor line of thought and practice carries along an insistence that we are not entirely ensconced in or wholly determined by interhuman relations alone and that other, more remarkable, ways of life are possible for us—especially if we learn to live more fully in view of our animal and more than human relations. The notion that resubjectification lies exclusively or even primarily within interhuman registers is seen by this minor tradition as an instance of a dogmatic anthropocentrism that inhibits genuine flourishing. In pursuing mental ethologies of the sort examined in this chapter, I am taking up and further developing this minor tradition in view of the specific anthropocentric systems of power that characterize our present.

Becoming Worthy of the Event

In closing this chapter, I should note that the account of resubjectification developed here must be carefully distinguished from the familiar gesture of emphasizing the primacy of personal virtues and individual character formation and responsibility in relation to social issues. This conservative approach often serves as a means of avoiding the political sphere and the difficult task of making structural changes in the established order. By contrast, the mental ethological framework sketched here is premised on the notion that addressing anthropocentrism requires fundamental structural transformation in all major spheres of life, from politics to economics to education and beyond. At the same time, it is important to note that it does not suffice simply to identify the pressing need for structural change in human-animal relations; it is equally essential to undertake the subjective changes required to initiate, implement, and sustain those changes.

If the analysis of anthropocentrism in this chapter has hit its mark, then we must acknowledge that the changes required to build a post-anthropocentric form of life are profound. Minor reforms in our patterns of consumption, extant policies, or institutional arrangements will not suffice to achieve such ends. We are faced with the need for a complete reordering of every aspect of our collective lives; and such thoroughgoing transformations are never achieved by intellectual commitments alone. They are driven and sustained by impassioned subjects who have glimpsed another way of life and who have a lived, existentially grounded commitment to actualizing that other mode of existence.

I have suggested in this chapter that, in regard to anthropocentrism in particular, such subjects must be built through experiments in mental ethology that displace our all too human perspectives and habits. Without undergoing these experiments in resubjectification with our animal kin, it will be difficult to appreciate the scope of the problems and challenges at hand. As we will see in the next two chapters, developing a more affirmative way of life with animals requires a thorough rethinking and reconstitution of our social and ecological relations as well. But the depth and extent of such changes and the necessity for undertaking them will remain hidden if we fail to undergo the requisite subjective reformation. This has been one of the great stumbling blocks of the contemporary animal justice movement, which often limits its focus to what it takes to be achievable, pragmatic reforms at the individual and political levels, without appreciating the full range of the problems that attend anthropocentrism. These problems can be held firmly in view only when they are approached from within the context of reconstituting relations with the beings—both human and more than human—who are most affected by those very problems.

CHAPTER TWO

Social Ethology

I must have been under surveillance for some time. For months I'd maintained an early morning routine of putting out water for the neighborhood squirrels and birds before walking to the gym to exercise. Over time I noticed that some of the birds were timing their visits to match my schedule. Most birds studied me from afar, waited for me to leave for the gym, then came to the water bowl to drink. One bird, though, went beyond just watching me from a distance and initiated an encounter. Much as in Joe Hutto's encounter with Rayme (see chapter 1), I was chosen by a particular bird: an American crow (*Corvus brachyrhynchos*). He was singular among the other crows (with a distinctive call and striking patches of white on his breast) and bolder than the other birds in the neighborhood, as crows often are. But even among his fellow crows, this bird was remarkably gregarious. He followed my movements intently, flew closer to me than other birds dared to, and seemed to find me and my family of great interest.

We started to watch him with equal interest. We called him Patches because of his distinctive white markings, which made it easy to identify him and track his comings and goings throughout the day. One afternoon Patches decided to do something no other crow has ever done with us: he perched on our porch railing—a vulnerable place for a large bird, since it is just a few feet from our living room window and the front door—and sat for several minutes watching us through the glass. Then he stretched out, fluffed himself up, and emitted a loud and very distinct three-count "caw" while staring straight at us. After this he relaxed into his resting position and waited for us to respond.

We thought he might be thirsty, but the water bowl was full. Then it occurred to us that he might be hungry. We typically avoided putting out food because in the past several close neighbors had complained about the birds it attracts (crows, in particular). We knew, though, that this was nesting season and he might need food for his mate and nestlings. So we put some peanuts out beyond the porch and waited to see what would happen. Instead of eating the nuts, Patches took as many as he could fit in his beak and flew out of sight.

We began tracking his flights and found that he was indeed feeding a mate and nestlings high in a pine tree about fifty yards from our house. On his rounds each day, he would come to the porch for a few nuts, sometimes soaking them in water, and take them to the nest. We held our breath over the next few weeks, waiting for the fledglings to appear. Finally Patches showed up on our porch with his partner and two fledglings in tow. The fledglings were slightly smaller than the adults, flew awkwardly, and made unusual calls. They still relied on the older birds to feed them. American crows are loud and obtrusive birds, and the fledglings were even louder. Their cries for food were incessant and piercing, and they decided to make the area we share with our immediate neighbors their home. They began waking us at an ungodly hour and established an increased presence in the neighborhood. We held our breath that our neighbors wouldn't complain.

Eventually the fledglings matured and could feed themselves. They made less noise and grew to be indistinguishable from the adult birds. As the weeks passed, the little family of four crows we'd become familiar with grew into a larger murder of seven and sometimes eight crows. This group has become a permanent fixture in our neighborhood and in our lives. They still visit us every day, sometimes just to check in and drink some water, sometimes for a snack. Despite daily contact, all members of the group stay a considerable distance from us except for Patches. He still shows up daily, lands on the porch railing a few feet from our door, and makes his usual request for food. His role has shifted from feeding his offspring to looking out for the entire group. If we toss out nuts for him and the other birds, he almost always waits until the others arrive and lets them feed first.

For me and my family, these crows are now kin; they play a prominent role in our lives, as we do in theirs. They spend hours resting in the trees and foraging in the grass. They are intimately familiar with our comings and goings and keep a constant eye on us. If a coyote passes through our neighborhood, they let us know. Patches occasionally leaves small items—gifts, perhaps—on our porch.[1] As our kin, the crows figure in any number of our daily doings and in our long-term decisions. We worry about them frequently and in manifold ways. If we leave for a few days, we try to find someone else to give them fresh water. When we've considered moving to another part of town, we've worried about breaking our kinship bonds. When the crows dig up the grass to find bugs and other food, we worry about which pesticides the landscapers use. We also worry that our contributing to the relative survival and reproduction of this species may be crowding out other bird species in our region and elsewhere. In caring about the crows and striving to be good friends to them, we thus find ourselves implicated in a number of complex social questions. How best, we often wonder, should we negotiate these social relations?

Crow *Culture*?

Before pursuing this question, and to preview the main points of this chapter, let's pause to consider some basics of crow ethology, especially crow sociality. Scientists have long assumed that, because of their relatively small brains, crows and other birds are less intelligent than bigger-brained animals. Recent research has demonstrated, however, that crow intelligence in particular is on a par with primate intelligence in various ways.[2] Not only do crows regularly use tools, but they have been shown to engage in novel *compound* tool use.[3] In a recent experiment, several New Caledonian crows were able to retrieve food hidden inside a box when given several sticks that individually were too short to reach the food. Without instruction or guidance, some of the crows quickly assembled the sticks into a new tool that was long enough to retrieve the food. This indicated that the birds were not simply using trial and error but could form a mental picture of a novel tool before it actually existed and could predict that it would work.[4]

Although the underlying cognitive processes involved in this advanced tool use are not well understood, such a high level of intellect among crows indicates sophisticated conceptual capacities. Researchers who study crow communication have identified nearly two dozen distinct caws that crows make, ranging from assembly and dispersal calls to warning and courting calls.[5] Anyone who spends time listening to and watching crows knows that caws that sound identical to humans are often accompanied by gestures and postures that produce starkly different responses in other crows. Similarly, crows will often sit near one another and "ramble," making a series of indistinguishable sounds (researchers sometimes call this form of communication "cooing" or "gurgling") that seem to vary with context. Crows also mimic sounds of all sorts (from humans, animals, even machines), sometimes apparently to get food or something else they need. (A crow in our neighborhood seems to imitate a cat—for what reason, we have yet to discern.) Crow roosts are notoriously noisy, and it's hard to believe this is merely idle chatter. Researchers believe that roosts and other group gatherings are opportunities to pass along essential information about food sources, dangers, and other salient matters.

Learning to avoid danger might be one of the reasons behind another remarkable phenomenon associated with crows: funerals. In crow funerals, typically an unusually large group gathers noisily at an unusual time on the ground or in the trees near a dead conspecific, watching the body for a time, then silently departing en masse. Researchers are not sure of the purpose of these events. Perhaps the birds are marking the death of an important member of the group, since not every crow death is followed by a funeral; it is also possible that the birds are temporarily protecting and observing a bird they believe might wake up. One experiment by Kaeli Swift and John Marzluff demonstrates that funerals are sometimes a way for crows to warn each other of the danger that has killed a conspecific.[6] In this experiment, crows tended to avoid the site where they witnessed a dead bird for up to six weeks, even though food was regularly available there. Although I have never witnessed a crow funeral, on my daily bicycle rides it is my long-standing habit to remove dead crows from the road. While doing this, I am, almost without fail, watched intently by other crows in the area. Whatever significance death might hold for crows, it is evident that they are keenly sensitive to it and communicate with each other about it.

These sophisticated capacities and collective behaviors all point to sociality as central to crow life. Crows commonly live in groups of extended family members who work together to locate and share food and other resources; they also use cooperative breeding strategies to feed fledglings and incubating females.[7] And though not all corvids are cooperative breeders, researchers have recently found that prosocial and generous behavior is more common among such groups, with more dominant birds giving others in their group priority in securing food and other resources.[8] I have frequently witnessed what appear to be both cooperative breeding strategies and prosocial behaviors among the crows in my neighborhood; in particular, Patches and other established members of his murder seem to be looking out for several birds in their kin group by fighting off attackers, ensuring that food is appropriately distributed, and so on.[9]

Given the extent of social learning among crows, are we warranted in going further and suggesting that crows can be said to have culture? Culture has, of course, long been seen as an exclusive property of human beings. Animals have commonly been characterized as purely instinctual, mechanical, reactive beings, incapable of learning, adapting to new circumstances, or adopting alternative means of communication, and hence as existing outside the scope of culture proper. This standard denial of animal culture, while still common in the humanities and social sciences, has to a significant extent been abandoned by ethologists who carefully study long-lived social animals. The evidence amassed for the existence of culture among such animals—especially cetaceans, primates, and many kinds of birds—is highly persuasive, even if it is not universally accepted by scientists.[10]

Leading crow ethologist John Marzluff argues convincingly and at length that crows and other corvids show clear signs of having culture.[11] But instead of simply listing the ways certain crow behaviors (such as the local transmission of tool-use techniques, communication strategies, identification of individual predators and harassers, and so on) indicate cultural transmission, Marzluff suggests that crow culture is best viewed and appreciated from the perspective of the way crow and human cultures have coevolved. In biological terms, coevolution refers to the ways two species affect each other's evolution through natural selection. *Cultural* coevolution is an extension of this idea into the cultural realm and is meant to denote the ways the cultures of two species affect each other's development and survival. Along these lines, Marzluff notes that human beings and crows have lived side by side for millennia and that crows have been viewed in various ways in different cultures: from being sacred relations worthy of protection to being pests and competitors deserving extermination. Natural-cultural patterns of interaction have changed based on these varying views, allowing crows more or less of a place among human societies (for example, by intentionally providing food for them or by trying to exterminate them through hunting).

Cultural coevolution sometimes takes place on rapid time scales as well. Crows have, for example, quickly learned to thrive in certain urban areas where food is easily accessible and many of their competitors and predators have been removed. The individuals and groups that flourish in such environments are successful in acquiring and transmitting information about which foods, water sources, and locations are safe, which kinds of human and domestic interactions are beneficial or dangerous, and how to find new roosting and nesting sites. Such cultural-environmental conditions are undoubtedly among the most important coevolutionary pressures on crows at present. In terms of the reproductive success of individual crows and groups of crows, it might be that bolder and more charming crows like Patches will be more successful in certain locales, whereas similarly gregarious behavior might be shunned in different cultural-environmental circumstances. In these and other ways, it is clear that modern human societies are pressuring crows to become ever more intelligent and discriminating—and that we are being forced, in turn, to rethink the status of these birds in our societies. Where do animals like crows fit in our social taxonomies? If they are not fully domesticated, can we say they belong to our societies? If they are not fully wild and have evolved alongside us for millennia, should we nevertheless try to limit our interactions and aim to treat them as if they were wild? And can we still have confidence in such categories as "domesticated" and "wild" in an age where widespread ecological and social changes have fundamentally overturned traditional ideas about where culture ends and nature begins?

What Is Social Ethology?

These are some of the questions posed by what I call *social ethology*. In the introduction, I explained that the term *ēthos* has a wide range of meanings that can be exploited in order to reframe and rethink human-animal relations. One of the primary meanings of *ēthos* is *custom*, in the sense of the shared practices and relations that constitute a given social order. With this sense of *ēthos* in mind, let us initially define social ethology as the careful study and consideration of what constitutes social life (1) among animals as well as (2) between human beings and animals. Given what I noted in the previous section about the ongoing coevolution of certain human and animal societies, social ethology obviously cannot be limited to objective reporting on existing social groupings and cross-species interactions; it must also rethink how a society is formed, unformed, and re-formed across time. At stake in the practice of social ethology, then, is a reconsideration of the very nature of the *social bond*—a term that should be understood to include, among other links and ties, companionship, kinship, friendship, association, alliance, assemblage, family, and community—as it is enacted and transformed among animals and between and among human beings and animals.

Dominant conceptions of social life tend to emphasize groupings with clear boundaries and fixed territorial markers such as the family unit and the state. The persistence and attractiveness of such well-defined social groupings is that they allow us to be certain about where relations and responsibilities begin and end. On the standard view, the boundaries of such groupings can subsequently be revised and redrawn in view of what are believed to be inadequate or unjust exclusions and delimitations (for example, we might acknowledge an expanded notion of the family to allow for new family configurations, or adopt a more expansive configuration of the state with more porous borders in order to accommodate new citizens or territories); but this view tends to proceed on the assumption that communities *must* have such boundaries and markers and that we are (at least ideally and in principle) in the process of appropriately adjusting them to match preexisting realities and conceptually consistent criteria for inclusion. Indeed, one of the reasons contemporary animal liberationist arguments have had such persuasive logical force is that they are able to demonstrate rigorously how widely accepted criteria for delimiting community membership (for example, sentience or subjectivity) logically entail including animals previously excluded from consideration, thereby providing rational justification for projects of pro-animal social reform.

But what if we went beyond this approach (admirable, to be sure) of trying to extend and reform extant social orders according to preexisting principles and categories of relation? What if, instead of taking the extant social order and its governing principles and categories largely for granted, we sought to conceive of the social bond as something that is inherently open-ended—as something that emerges in and through experimenting with new relations that undo old configurations and open up different potentials and possibilities? A social ontology that begins from the latter premises would not aim primarily to track preexisting lines of inclusion and exclusion but would instead be attuned to the relations that emerge in, around, and between such delimitations as well as to the alternative groupings that might form if other sorts of encounters and configurations were pursued. Such a perspective would thus begin from the premise that we should hold open the possibility of rethinking the social bond at all levels, from microrelations within a given body to macrorelations at the intersubjective, ecological, planetary, and cosmic levels. In this approach, there is no preexisting way to cut the social world apart at its joints, no assured way to discern some kind of rational or inherent "grain" to our social worlds. The only way to know what constitutes a social bond is to try it—to experiment with a given relation and see what configurations of social life emerge from it.

Social ethology as I have sketched it here should be understood, then, as a creative act, as enacting alternative modes of sociality: being-with, being-together, and other such affective relations. And although the sorts of intersubjective and social relations I have canvassed thus far with deer or crows would likely not be replicated with, say, certain predator species, this does not mean that sociality is somehow less plastic in such species or that relationship always entails the same kind of intimacy.[12] Although the variables and potentials that give rise to a particular social grouping might be very different from species to species, the wide variety of relationships that have been established with many of the animal species human beings have encountered (relationships that range from violent domestication to pacific coexistence), as well as the astonishing variety of interspecies relations beyond humanity, underscore that social agency and the potential for alternative forms of sociality are widespread among both human and nonhuman beings. How far these potentials might extend, and what alternative configurations of social life might emerge from them, cannot be fully known in advance.

As I noted in the previous chapter, one of the defining traits of a deep ethological approach is the willingness to undergo deterritorialization, in the sense of both a subjective and a topological change of social life. Deterritorialization in this sense does not necessarily mean traveling to a new location (although such movements are not ruled out); rather, the point is that a social ethological disposition inflects one's inhabitation along new social-relational lines, thereby opening up new modes of knowledge and new possibilities for living. From this perspective, social ethology can be understood as inviting us to practice a kind of "gay science," one that moves beyond a notion of knowledge aimed at bringing what is strange back to oneself and that gestures instead toward a form of inquiry and relationship predicated on gratitude and on reverence for what is inhuman and more than human.[13]

These sorts of social experiments, though, raise profound questions about the places we inhabit. Even if gay science and social ethology do not necessarily entail leaving one's home or city, they certainly entail transforming one's abode in the direction of more respectful and more experimental encounters with our more than human kin. The central challenge of social ethology is thus to redirect our societies away from their all too human focus and to transform them into thriving multispecies communities.

Animals in the Polis

Félix Guattari (from whom, recall, I borrowed the tripartite schema for the main chapters of this book) offers us a useful starting point for considering how one might undertake such social transformation. When Guattari turns in his own work to the question of what degrades social relations, he raises many of the issues one might expect given his radical political disposition, issues ranging from homelessness and the subjection of women to social intolerance and urban decay. What is striking (though not altogether surprising) is his overarching emphasis on the role that integrated world capitalism (a synonym for what is more commonly referred to today as postindustrial capitalism or neoliberal globalization) plays in the constitution and critical limitation of contemporary social formations.[14] Even if for Guattari social injustice does not reduce simply to economic inequality, his analysis grants capitalist economic relations clear priority for understanding how injustices arise and for assessing where obstacles to radical social transformation are situated.

Whether or not one agrees with Guattari's neo-Marxist position on the contemporary degradation of the social bond, one can hardly deny that economic arrangements have an inordinate influence on social relations, both human and more than human. It might seem to follow, then, that my analysis of the reconstitution of sociality might track in tandem with Guattari's and lead us to examine how human-animal and animal-animal relations have been both reconfigured and degraded by capitalist economic relations in the form of the "animal-industrial complex," the joint exploitation of human and animal labor, and related human-animal economic assemblages.[15] Without denying the importance of capitalist economic relations in forming this problematic, I want instead to pursue a broader approach to the issue of sociality in order to avoid adopting too quickly (as many Marxist and neo-Marxist theorists, Guattari included, risk doing) a previously delimited notion of social relations. For what is at stake in doing social ethology, as I just noted, is a thorough reconsideration of the very nature and scope of the social bond. Thus, even if we were to agree that economic relations are among the most important factors in the constitution of contemporary social life (and I should underscore that I do believe they are), this stance would not allow us to raise in a fundamental way the question of the nature and scope of the social bond. How we answer that question determines the very matter for thought in the domain of the social, a matter that critical economic analysis often bypasses in order to get under way. Thus, beyond the economic question lie a series of more basic and more fundamental socio-ontological concerns: Who or what is my neighbor? Who or what belongs to the polis and to the household? What is the extent of kinship? Is it possible to establish friendships and political relations with animals?

These socio-ontological questions are by no means new. They have a long history in Western culture, and the answers provided to them mark a split in the political foundations of that tradition between (1) a dominant approach that denies justice and the extension of community to animals and nonhuman life, and (2) a minor approach that sees animals as kin, capable of friendship, and deserving of justice and respect. The first, dominant tradition is given voice early on by Hesiod, who tells us that only humans have a sense of justice and that righteousness is absent from the hearts of animals.[16] This basic posture is further articulated and defended by such prominent philosophical schools and figures as the Sophists, Aristotle, the Stoics, and the bulk of medieval, Renaissance, and modern philosophers from Augustine and Thomas Aquinas to René Descartes and Gottfried Wilhelm Leibniz and to Immanuel Kant and Georg Wilhelm Friedrich Hegel.[17] The second, minor tradition can be found in Pythagorean and related schools that stress human-animal kinship and continuity and often advocate for individual and collective practices that extend considerations of justice to animal life. Elements and refinements of these ideas recur among Empedocles, the ancient Cynics, Lucretius, pro-animal philosophers such as Plutarch and Porphyry, and a variety of rogue philosophical figures including Michel de Montaigne, Jean-Jacques Rousseau, Mary Wollstonecraft, Arthur Schopenhauer, Henry David Thoreau, Margaret Fuller, Friedrich Nietzsche, and Henry Salt.[18] The social ethological approach I am developing here belongs squarely within this second, minor tradition and seeks to recover and pursue the questions it poses.[19]

In adopting an explicitly socio-ontological and political framework for the issues at hand, I am also diverging from the approach to animal issues that has been most common among pro-animal philosophers and theorists in recent decades. For most philosophers, the question concerning animals has been seen primarily as an *ethical* question, in the sense of whether and to what extent we as human beings have ethical obligations toward animals. Posing "the animal question" in this particular way has led to a largely individualist (What obligations do human individuals have to individual animals?) and abstentionist (Which animal-harming practices should human individuals abstain from?) approach to animal ethics, with much of the discussion revolving around ways human beings can use their individual agency to reorient markets and institutions in a more animal-friendly direction.

In response to limitations with this approach, animal philosophers and animal studies theorists have undertaken a more direct and explicit "political turn" in their work. This turn is necessary, they argue, because the ethical approach to animal issues—with its individualist and abstentionist emphasis—fails to get at the underlying structural factors that give rise to injustice and violence directed at animals. For when individual decision-making is the focus of critical analysis (as is typically the case with mainstream animal ethics), it is often difficult to see the deeper collective, social milieu in which decisions are made. By reflecting more carefully on that social milieu—which includes everything from economic arrangements to the educational, legal, media, and legislative infrastructures that shape our collective thought and practice—it becomes possible to consider the more fundamental questions of how the boundaries of society, community, and kin have been drawn and how those delimitations have shaped infrastructures as well as the dispositions of individual decision-makers.

Among the foremost proponents of the contemporary political turn in animal studies are Sue Donaldson and Will Kymlicka. In their influential work *Zoopolis*, they argue that the mainstream animal ethics strategy of aiming chiefly to prohibit violence against animals is insufficient for social transformation, because such an approach fails to acknowledge the irreducibility and unavoidability of human-animal interactions.[20] Venturing beyond an individualist ethic of noninterference, Donaldson and Kymlicka develop a political theoretical framework that aims to rethink a number of human-animal institutions and interactions. What is perhaps most novel about their approach is its pluralism. Rather than proposing a single criterion of normative mattering that frames animals in homogenous terms and a single framework for governing interactions (monistic tendencies that are all too common in mainstream animal ethics), they recognize that human-animal relations and institutions take various forms and manifest varying levels of affect and influence. Any political framework that seeks to do justice to these relations, they argue, needs to be correspondingly plural and flexible.

Thus, even as Donaldson and Kymlicka share with animal rights approaches an emphasis on the importance of respecting individual animal lives, they recognize that justice for individual animals will entail different postures and practices depending on the individual animal or the species of individual under consideration. Along these lines, they recommend adopting a threefold framework for reconfiguring interactions with animals that takes into account the multiplicity of animal lives and relations. Domesticated animals, whom we live and interact with in regular and intimate ways, they classify as full citizens. Wild animals, whom most urban dwellers remain at a distance from, they consider members of independent sovereign communities. Liminal animals, such as the undomesticated crows I have been discussing in this chapter, who live in and on the edges of urban locales, are characterized as denizens having different sorts of rights depending on their status and interactions with people.

This threefold framework is meant to be not taxonomically exhaustive but pragmatically effective within the existing categories and institutions of citizenship theory and liberalism (which, as Donaldson and Kymlicka rightly note, is the dominant political framework in many contemporary nation-states). Liberalism, they maintain, has a series of ready-made concepts, traditions, and institutional arrangements for dealing with deep differences, and they recommend putting these tools to work on human-animal relations. Although Donaldson and Kymlicka's framework is unabashedly reformist, we should not assume that their approach entails only minor changes to dominant ways of life. They make it clear that adhering to even the most basic tenets of citizenship, sovereignty, and denizenship for animals involves thoroughgoing changes to a wide range of structures and institutions: not just foodways and invasive experimental procedures (as mainstream animal ethicists tend to emphasize) but also, for example, systems of mobility and urban infrastructure (registers of social life that are all too often overlooked in pro-animal discourse and activism). Thus, even for more radical pro-animal advocates who do not share Donaldson and Kymlicka's commitment to liberalism, we must acknowledge that their approach allows us to pose the question of the place of animals in social life in several domains and institutions that mainstream ethical frameworks have traditionally neglected.

Ultimately, though, the key strengths of this framework—acknowledging the importance of the lives and deaths of individual animals within the sociopolitical domain, recognizing a variety of human-animal interactions, and making a pragmatic effort to work within the dominant ideological and institutional frameworks of contemporary liberal democracies—have certain shortcomings. First, despite the undeniable importance of liberalism's emphasis on the sociopolitical significance of individuals (something that must be underscored in regard to animals, who are often denied their singularity and subsumed within anonymous totalities by industrial forces), the ideological commitment to individualism is ill-suited to deal with the complex issues that surround animal collectives and assemblages. This is true in terms of kinship groupings as well as the complex interspecies relations that form the ecological context in which the bulk of animals live. This does not entail, of course, that we should adopt, say, an ecocentric approach over an individualist one when trying to determine our political obligations and relations with other animals; the point is that political life with animals entangles us in complex and irreducibly plural considerations in which respect for individual subjects is but one consideration among many (and often not the leading one).

Another problem with the sort of liberal animal politics Donaldson and Kymlicka espouse is its presumption that extant modes of human-animal relations are largely fixed—that is, their approach lacks the sort of experimental, deep ethological sensibilities we are exploring here. Donaldson and Kymlicka are entirely willing (and commendably so) to consider the political rights and standing of a range of animals—domestic, liminal, and wild, to use their categories—but they show little interest in experimenting with and accepting reconfigurations of the types of relations that have given rise to these categories. There is little room in their framework, for instance, for transforming urban life to allow for deeper experiments in living with wild animals or in experimenting with forms of life that might fundamentally alter our views of the nature of domestic and liminal animals. Rather, they seem to assume that the categories of relation that characterize contemporary urban life in modern liberal states reflect a realistic account of human and animal differences and that the primary political task is to adjudicate disputes and address injustices that occur within and among these (relatively fixed) categories of relation. Reading their work, one does not get the sense that liminal animals or wild animals might, like the crows in my neighborhood, become our kin and ask more of us than justice, or that relationships with animals currently deemed "domestic" might undergo unanticipated developments that challenge us to reconsider domestication as such. A social ethological approach, however, views the social domain as something always to be negotiated anew—not simply in view of making its current form more just, but also with an eye toward reconsidering its very nature and scope.[21]

Multispecies Relations in a Crowded World

Donna Haraway's work serves as a helpful corrective to these problems with Donaldson and Kymlicka's framework and stands as an important contribution to animal politics in its own right.[22] Like Donaldson and Kymlicka, Haraway refuses the notion that animal politics primarily entail an ethic of noninterference and avoidance. Instead, she insists that human-animal relations and mutual affectivity are ubiquitous, irreducible, and unavoidable. What is more, Haraway understands such cross-species relations as something to be affirmed, as a good part of the basic ontological fabric of earthly life. She suggests that, at the ontological level, relations go all the way down, and that individuals should be viewed as emerging but never completely severed from the relations that constitute and sustain them. As we will see, it is Haraway's commitment to a thick version of relationality that brings her work into the orbit of the philosophical ethological approach developed in this chapter and that allows her to avoid reifying categories of human-animal relations as Donaldson and Kymlicka tend to do.

Haraway develops her ontology of human-animal relations most influentially by way of the concept of *companion species*.[23] Let's examine both parts of this concept in turn. First, as I just noted, for Haraway companionship is basic to earthly existence. Our companions—whether human or more than human—consort and break bread with us (Haraway here reminds us of the etymology of the word companion, from the Latin *com-* [with] and *panis* [bread], meaning one who breaks bread with another); and we build meaningful worlds, lives, and societies in common. To share life with others, however, also means to share death with them. Relationships and companionships are always built among beings who are finite (born into a world not of their own making) and who are mortal (destined to die and decompose). To be a genuine companion, a true comrade, is to care about these life-and-death relations and to learn how better to inhabit them.

Haraway uses the second concept, *species*, to recall readers to the perspective and historical scale of biology and evolutionary theory. While commonsense notions tend to envision species as fixed, static, or natural kinds of some sort, Haraway argues that they should be understood as historical, dynamic, and relational in nature. In her account, species are characterized as groupings that emerge under specific historical and evolutionary pressures and as such are contingent (the groupings could have taken a different form had they evolved under different conditions and will of necessity become something else over time as relations and selective pressures change). So a species is itself a "multispecies crowd," internally differentiated and externally related to numerous species at micro- and macro-levels. Haraway also emphasizes the rich etymological senses of the word species, which range from wealth and filth to beholding and looking. This latter register of terms having to do with the act of looking (from the Latin, *specere* [to look at, observe]) are especially important for Haraway because they point toward the ethical dimension of companion species. To inhabit multispecies relations in a way that does justice to them and understands them as an instantiation of "the good" is to look thoughtfully and carefully at those relations—to hold them in regard and respect (*re + specere*).

As is evident from both components of the concept of companion species, for Haraway the social bond is capable of being made and remade in various ways, formed and reformed in and through varied and concrete interactions. Haraway's ontology thus gives us a vision of social life that is considerably more plastic and open to transformation than the vision operative in Donaldson and Kymlicka's work. Although Haraway recognizes something we might call different "categories" of relation between subjects, in her work such categories are far more susceptible to becoming undone and redone. Within a companion species ontology, for example, there is no guarantee that a certain domestic species might not become less domesticated in a particular context, or that a species thought to be irreducibly wild will not partially adopt domesticated behaviors or permanently invade domesticated spaces. In a certain sense, the concept of the liminal in Donaldson and Kymlicka applies to *all* beings in Haraway's schema, for both human beings and more than human others reside in threshold zones and occupy messy and disruptive assemblages at the intersection of what Haraway calls *naturecultures*.[24] This messier and more flexible conception of social relations, as we will see in the following chapter, better illuminates the increasingly crowded and complex world of radical ecological change we inhabit. Climate change, defaunation, and other forms of ecological degradation are poised to bring about—and have to some extent already brought about—a new world wherein traditional categories of relation (wild, domestic, and so on) are, for better or for worse, thrown wide open.

The task of social ethology is not to beat a retreat from this new reality but to take it on board and see how best and most respectfully to negotiate its potentials and perils. While Haraway's work is profoundly useful for elucidating a more expansive concept of the social bond and for articulating the complex relational realities of a "claustrophobic" world of ecological turbulence, it is decidedly less helpful in its attempt to negotiate the political implications of this expanded notion of social relations.[25] Generally speaking, Haraway's animal politics tend to lack the imagination and daring of her ontological forays. To illustrate, consider one of Haraway's favorite examples for epitomizing the plastic and messy nature of the social bond: dog agility training. On Haraway's analysis, this sport is an instance of a companion species relation in which the dog and human trainer work together to create an outcome that would be impossible if the two were entirely independent units. Dog agility training (at least when done well and successfully, in Haraway's view) is, then, not a simple act of a human being exerting power and dominion over a particular dog; instead, successful training requires that human beings carefully and thoughtfully negotiate with the heritage of a given dog species as well as develop a set of practices that foster play, joy, and creativity on the part of both dog and human trainer. Without this responsive companionship, dogs will not enjoy agility training or undertake it with gusto. Ultimately, Haraway suggests, training of this sort allows sporting dogs the maximum freedom possible within contemporary human societies—"the freedom to live safely in multispecies, urban and sub-urban environments with very little physical restraint and no corporal punishment while getting to play a demanding sport with every evidence of self-actualizing motivation."[26]

Despite Haraway's creative defense of dog training and its seemingly positive sociopolitical implications, many critics have rightly suggested that her approach fails to ask deeper questions about the nature and implications of such relationships with animals.[27] While it is undoubtedly true that domestication has historically been a two-way relationship between human beings and animals (and offers powerful evidence of the flexible and mutable nature of social bonds), there are good reasons for asking whether practices involving domesticated animals, like agility training, are worth preserving. What kind of people do we become when we think of ourselves as animals' "tamers" or "trainers"? What vision of animals and of our place in the world inheres in such a stance? Further, aren't there other ways of life that would allow domesticated animals to live rich and rewarding lives beyond such rigid training regimens and outside the confines of contemporary urban life? If our aim is to live justly with and alongside animals, might it not be preferable to work toward transforming urban and other human-centered environments to maximize the range of animal potentials rather than taking those environments for granted? And might it not be best to seek to abolish many of the practices associated with domesticated animals and focus instead on transforming the way of life that renders ongoing domestication necessary?

In much of her work, Haraway tends to overlook these more radical possibilities and instead assumes that our traditional human-animal institutional arrangements are largely fixed and worth preserving (and here she falls short of the more transformative political vision we find in Donaldson and Kymlicka). Along these lines, Haraway adopts a similar inevitabilist and reformist stance on the issues of eating animals and using them as experimental subjects. These practices, she maintains, have emerged and coevolved with animals over long periods, are not inherently objectionable, and can be reconstituted in ways that are more respectful of animal lives and deaths. Further, Haraway seems to think that if a particular practice is abolished (say, cattle farming or invasive experimentation), then the animals involved in that practice will either go extinct or become like museum exhibits.[28]

But if, in fact, the social bond can be unmade and remade in different ways—a point Haraway's work perhaps emphasizes more than anyone else's—then so might corresponding institutional and structural arrangements. Thus we might wonder, Isn't it possible for domesticated animals to be given the space to live differently in and among us? And need we continue eating and experimenting on animals simply because these are inherited practices, or because the dominant culture lacks the imagination to envision and enact alternatives? Need we accept the institutional arrangements of contemporary urban life as we currently find them and try to fit various species of animals into them? Why not proceed in the opposite direction and seek to rearrange social life with the lives and deaths of animals and other more than human beings as the main concern? Such questions should not be taken to imply that social bonds are infinitely plastic and that social arrangements can be changed haphazardly; but we must always bear in mind that human-animal relations and social formations are astonishingly diverse. Even rudimentary ethnographic awareness should remind us that human-animal relations take considerably different forms in a variety of social contexts and that urban/suburban forms of life, despite appearing relatively fixed and permanent, are in fact fundamentally contingent and subject to ongoing change.[29]

Reenchanting the Polis

I have reached a critical limit in my elaboration of social ethology. I have highlighted the need to fully reconsider the nature and scope of the social bond as well as to rethink the social and institutional arrangements that grow out of that bond; but my analysis of Donaldson and Kymlicka and of Haraway has demonstrated just how challenging it is to envision how the two registers might fit together and be jointly transformed. As a means of addressing this limit, let's return to the notion of a zoopolis examined earlier in the chapter. Recall that Donaldson and Kymlicka developed this concept in an effort to think more carefully about the nature of the social and political spaces we live in and about how animals are both included in and excluded from these spheres. Donaldson and Kymlicka offer a heuristic schema of domesticated, wild, and liminal animals to help us think about the relevance of various forms of animal life to human political institutions. Although their proposal for a zoopolis is fleshed out using categories of relation and strategies for political transformation inherited primarily from the liberal political tradition, they borrow the notion of zoopolis itself from an earlier work with that title by Jennifer Wolch.[30] In closing this chapter, I will return to Wolch's original essay, as her vision of how a zoopolis might be constituted proceeds very much in the same spirit as the social ethological perspective elaborated in this chapter and can also help us address the critical limit we encounter at this juncture in the analysis.

Wolch's approach begins from the premise that animals have their own sorts of subjectivity and that improvements in human-animal relations should be based on recognizing those subjectivities. In her analysis, Wolch further emphasizes that the maturation and integrity of human individuals and communities requires recognizing our subjective and ontological kinship with animals. She suggests that this recognition has been present throughout much of human history but has gone missing in certain parts of the contemporary globalized world. She also argues, as I have done here, that subjective transformation is not independent of social transformation. Recognizing the depths and effects of human-animal interactions forces us to grapple with the ways the established order has failed to respect such relations and integrate them into its basic fabric; and this failure leads us in turn to look to alternative forms of both human and more than human community for possible ways of reform and radical transformation. Thus Wolch does not suggest that mere intellectual recognition of animal subjectivity or of ontological kinship among human beings and animals will be enough to effect social change. Such recognition must be coupled with a thorough rethinking and reconstitution of our entire way of life, beginning with the structural and material environments we inhabit. The question, of course, is how to move from an abstract *recognition* of kinship to the *enactment* of kinship relations that allow for our joint flourishing.

Trained as a critical urban geographer, Wolch would have us think about such transformations in terms that are very concrete: material, structural, and infrastructural. In other words, transforming the kinship relations of the human inhabitants of modern urban milieus entails changing the urban environments that structure those milieus. From the critical geographical perspective Wolch develops, urban environments are to be understood as both a reflection of the anthropocentrism of their inhabitants and as one of the chief means whereby anthropocentric ideologies and material realities are reinforced and intensified. The modern Western schism between human beings and animals, she suggests, is both constituted and deepened in large part by the geographical arrangements of many of our cities and urban centers. Urban and suburban ways of life separate human beings from animals physically and architecturally through such means as road construction, fencing, home building, and landscaping. In adopting these ways of life, we remove countless animals from our purview and cut off the possibility of establishing alternative modes of relating with them. Further, many cities do more than simply wall human inhabitants off from animal others; their mode of operation is premised on the instrumental use of animals and on inflicting harm and death on them. In many ways, our contemporary suburban habitats function as "killer cities," structurally founded on the violent use and death of animals.[31]

Wolch's call to develop a genuine zoopolis—a polis that is welcoming to animals of all sorts—thus involves two significant challenges. First, if we are to make our cities more animal friendly, we have to reconsider a wide range of standard aspects of daily urban life. This would include, she notes, not only "animal regulation and control practices, but landscaping, development rates and design, roadway and transportation decisions, use of energy, industrial toxics, and bioengineering—in short, all practices that impact animals and nature in its diverse forms."[32] Thus, as Wolch implies here, a zoopolis that places animal justice and kinship in the foreground will not be achieved with a few minor reforms. Taking the social and political dimensions of animal lives seriously calls for reorienting our individual and collective ways of life at a fundamental, (infra)structural level.

But even implementing such (infra)structural changes (which, note, Donaldson and Kymlicka also call for) is insufficient according to Wolch. What makes her approach distinctive and brings it into line with the social ethological disposition of this chapter is her call to "renaturalize" urban space and to "invite the animals back in . . . [and] re-enchant the city."[33] This call for re-enchantment goes well beyond the modification of harmful structures and aims at a complete transformation of urban life. Here the goal would be to create living environments that make animals—and animal encounters—part of the warp and woof of everyday life. It is precisely this provocative suggestion that makes Donaldson and Kymlicka bristle. They readily acknowledge their debt to Wolch's zoopolis concept but are categorically opposed to the suggestion that undomesticated animals should be given fuller access to the human polis.[34] It is ultimately this issue that is at stake in our discussion of social ethology. Should we content ourselves with trying to fit animals into our existing urban lives and our political traditions and institutions? Or should we instead seek to rebuild our lives and institutions from the ground up in response to the dynamic interspecific relations and more than human world we find ourselves in?

The social ethological approach to human-animal relations I have sought to develop here predominantly takes the second line of approach and thus asks more of us than reforming existing institutions and extending ethical and political criteria to new groups of beings. It invites us to experiment with reconstituting the social bond, and it summons us to a more expansive narrative about ourselves and our planetary kin: a different story about who we are, what we might become, and how we might otherwise find our place in the world. As potentially broad and inclusive as Donaldson and Kymlicka's version of political liberalism might be, the story it belongs to and seeks to call us into is not rich enough to do justice to the dynamic relations and potentials of planetary existence. We owe it to ourselves and our more than human kin to try to tell a better story about social life; more important, we owe it to them to live in such a way that we seek to do justice to the sociality thus described. This sort of transformation can be enacted only by profoundly transforming the social bond itself and the social milieus we live in; and it must be grounded in what Wolch calls the local, embodied knowledge that is won from concrete experiments in re-enchantment.

To be sure, reenchanting the polis must be done with care and foresight concerning the risks such a project invites. To reconfigure urban life along these lines carries health risks, brings new threats to human beings and the animals who already occupy these social milieus, and, as Eva Giraud has sagely noted, might create new forms of exclusion.[35] The point, then, is not to assume that all forms of re-enchantment will be mutually beneficial and will necessarily lead to the flourishing of both humans and animals. Yet we must acknowledge that urban life at present has been structured to radically limit such relations and possibilities for living differently. There is ample room to undertake experiments aimed in the other direction: establishing maximally respectful relations. We must further grant that alternative ways of living with animals are not unprecedented in human cultures (on the contrary, our current form of urban life is rather exceptional) and that experiments in living alongside animals have already been enacted to various degrees and in various modes in communities across the globe.

Deep ethological forms of social living, then, need not be created entirely from scratch. Rather, the task of reenchanting the polis amounts to sustaining and refining a way of thinking and living that has multimillennial roots, with variants found both in the West and across several other cultures. These other traditions have sought to share life with animals and more than human others, to acknowledge profound ontological continuity and kinship, to explore genuine companionship and friendship with animals through and across differences, and to understand justice and the social bond as encompassing both human and animal existence. The impetus to engage in this task, though, ultimately derives from animals themselves and their potential for living different and more remarkable lives. Social ethology should thus ultimately be understood as responding to an invitation to experiment in building a worthwhile life in common.

CHAPTER THREE

Environmental Ethology

Linda Hogan's *The Radiant Lives of Animals* recounts a life spent dwelling with and caring for animals in the Rocky Mountains of Idledale, Colorado.[1] As readers learn, the relations Hogan forms with animals are inseparable from the land where those relations inhere. Land—the focus of the present chapter—refers here to the specific bioregion and ecosystems within which Hogan lives and writes, but the term has a broader significance in her work than can be captured by ecological discourse. Very much like the animals whose kin Hogan has become, the land itself must be understood as a complex, multilayered set of agents and forces that elude complete cognitive capture. That Hogan relates her story through a combination of poetry and prose and a range of literary genres—from myth and science to autobiography and narrative—is telling: it suggests that if we are to understand something of the significance of land and land-animal relations, we must be prepared to think, see, and feel in a variety of discursive and nondiscursive registers.

Hogan did not grow up in Idledale. With roots in both Colorado and Oklahoma, Hogan (who is Chickasaw) lived a nomadic childhood, following her military father on his deployments overseas. As an adult, Hogan took up residence in many locations to teach, write, and study the land and animal life. Her writings—which now span four decades and include novels, poetry, essays, and other nonfiction—reflect these varied existential climates and evince a consistently deep ethological disposition. Hogan's long-term sojourn in Idledale, though, belongs to a different sort of ethological experiment and derives from a very different environmental sensibility. These changes reflect her choice to remain for the past several years in a specific locale, which has allowed her to become more familiar with its inhabitants and care for them more fully.

Hogan's poetry and prose in *Radiant Lives* highlight dozens of moving relationships she has formed with animals, ranging through and across all the registers of animal life considered in the previous chapter. Hogan relates stories of her close relationships with horses and dogs, encounters with wolves and snakes, and unexpected visits from crows and elk. While Hogan respects the differences between these forms of animal life, she also realizes that such species demarcations by no means denote invariant kinds or groupings. Further, she is keenly aware that her disposition toward the animals she interacts with has the potential to change their disposition toward her. One of the refrains throughout Hogan's writings is the wish for human beings to become something other than the enemies of animals. The counter-ideal that animates her work is for human beings to live in "so beautiful and whole" (117) a manner that animals and the more than human world could come (once again) to admire, and even envy, us.[2]

Through the long-term process of staying with the land, observing its inhabitants attentively, and tending the land respectfully, Hogan has undergone a transformation in her sense of self. What I called processes of resubjectification in chapter 1, Hogan refers to as being *re-minded*, a term that involves enacting new mental habits, modes of perception, and "new ways of thinking about our shared world" (7). Hogan likens this particular practice of being re-minded to learning a new language, as it requires us to think about animals, land, and ecological relations in ways that cut across the anthropocentric concepts and discursive patterns of contemporary culture. Viewing the world with a mental ethological or re-minded disposition, Hogan suggests, is less a matter of mastery and more one of *beholding*, a kind of deep seeing in which we gradually catch fuller sight of the underlying fabric of our surrounding world and the patterns and connections that bind and loosen relations.[3]

For Hogan, learning to see in this manner is not something to be attained on one's own, nor need it be built entirely anew each generation. Discerning and studying the patterns and connections that inhere in the more than human world is, of course, something regularly practiced by ecologists and naturalists; and any serious effort to behold and care for the land and for animals will of course be indebted to this body of accumulated scientific research. But such practices of beholding are also a part of many long-standing Indigenous cultures. Indigenous ways of seeing and being in the world are particularly important for Hogan's analysis, as they often include the scientific knowledge acquired through ecological and naturalist study but go beyond a scientific perspective to include a broader set of relations, stories, and values that are immanent in the land and its inhabitants. For Hogan, it is this combination of scientific-factual and relational-normative beholding that provides an alternative vision of the world and our place in it and that offers us the possibility of pursuing a worthwhile way of life even under increasingly degraded environmental conditions.

Hogan emphasizes that respectful beholding is ultimately a matter of learning to be appropriated and owned by the land (13), of learning to see oneself as a guest and a visitor to an ongoing story that unfolds in and through the land. In ecocentric terms, the aim of re-minding oneself and beholding the world is to see oneself as a "plain member and citizen" of the ecology and planetary relations;[4] in Hogan's terms, it is to see oneself as belonging to a broader "wilderness congress" (14). To relinquish the role of master or manager of the land and to accept and affirm being owned and appropriated by it is to appreciate "the smallness of our human self and the greatness of the world around us" (26). It also involves recognizing that we human beings are not the only creative agents on the scene, for the land itself teems with various agential powers.

To recognize the sublimity of the land is not, however, to deny its vulnerability. To be sure, the dynamic land relations and earth systems we belong to have preceded us by billions of years and will outlast us by billions more; nevertheless, we can do inordinate and lasting damage to our planetary kin while we are here. Hogan's insistence on downplaying human significance should not, then, be read as an attempt to unburden human beings of their responsibility for the state of planetary affairs. Throughout her work she emphasizes the seriousness and ubiquity of ecological degradation and the need for reparative and restorative efforts along these lines. Hogan does not suggest that such restorative work will somehow entirely reverse current patterns of destruction and save all life-forms and processes under threat. But her work is premised on the notion that concerted efforts in this direction can certainly limit harm while prefiguring the world we want to help bring into existence. Thus the practices of re-minding and beholding are ultimately about recognizing our limited planetary and cosmic significance while at the same time exercising the agency we do have to be the kinds of beings who could be welcomed and admired by animals and the land.

For Hogan, it is Indigenous traditions and languages in particular that carry and preserve a normative understanding of animals and land. She notes that in her traditional Chickasaw language the word for animal (*Nan okcha*) means "all alive," a notion that goes beyond an animal's being merely animated to include sentience, subjectivity, and living within meaningful relational networks; in short, to refer to a being as an animal is to underscore its "being a significant part of a whole" (80). Animals, then, are understood to be entities who have relationships and connections of their own, oftentimes entirely independent of human life and concerns. Hogan further notes that awareness of such animal independence gives rise within her tradition both to a sense of ontological awe and to a feeling of ethical obligation; we are called upon to attend to animal aliveness and to do what we can not to destroy it, but to actively help sustain it. Enacting this ontological and normative vision is, for Hogan, "part of our human purpose here," a moral debt we owe to the source of our own lives and a way of affirming the shared condition of living beings (80). It also forms an aesthetic practice, in the sense of attending to the beauty and alterity of the more than human world. With this aesthetic, our interest and attention are directed toward animals' own worlds and relations instead of viewing them solely in terms of how they fit into our lives and projects. By learning to slide into and persist within this inhuman, more than human zone, we undertake the transformative discipline of decentering ourselves and our own perspective in order to behold a world beyond the familiar coordinates of everyday interhuman existence.

Hogan describes this sensibility in narrating an encounter with a fox she has lived near for some time. She explains how she has sought over time to establish trust with the fox, to form at least one relationship where "no cruelty has found its way in" (98). For Hogan, beauty in its truest and most profound sense is to be found where we witness and recognize a life, like that of this fox, that has no need of us. "[The fox] knows I have crossed this bridge. An ear is turned in my direction, listening to the weight of air as I send her a message that this human is harmless. I only wanted to pass this threshold between species, wishing to observe the beauty of other lives around me, the beauty of her life. It may be irresponsible but I want to see and know the lives that have no need for me" (98).

The ontological, ethical, and aesthetic framework Hogan sketches here, while heavily indebted to her own and other Native American traditions, is not, she stresses, the exclusive preserve of Indigenous cultures. Hogan reminds us that we find evidence of similar visions at work in the cave paintings of ancient peoples (with their painstakingly and beautifully drawn depictions of animal life and their decentering of human significance [47]); we also find traces of such sensibilities in our own time in minor cultures across the globe that work to restore ecological milieus and habitats and strive for the shared flourishing of human and more than human lives. Thus, even as Hogan emphasizes the specificity of her conception of beholding as emerging from a particular tradition and place, the general principles of respect for the land and care for the more than human world, she suggests, belong to all people. What is more, if in fact one desires that all relationships flourish, such beholding cannot remain the preserve of a handful of peoples but *must* become more widely adopted—for any small-scale efforts to establish and maintain such relationships are sure to be swamped by the broader anthropocentric established order. It is here, in the recognition for the need to multiply (if not universalize) the practice of beholding, that she makes the call to her non-Indigenous readers. For such readers, Hogan is not suggesting either appropriation or uncritical acceptance of a given Indigenous tradition but is instead inviting readers of all sorts to learn from a complex and ongoing set of Indigenous heritages and ways of life, in order to uncover resonances with those traditions and resources for transforming their own lives and worlds.

Ecological Degradation I: Extinction

Hogan's reflections have introduced us to the challenging but essential task of learning to attend to animals in relation to their environments, where environment includes both the ecological milieus animals exist in as well as land in the broader sense of the term. Let's take each aspect of this *environmental ethology*—a term for which we are slowly developing a sensibility and a definition—in turn, beginning with the ecological realities of contemporary animal life.

For those concerned about the fate of animal life today, coming to grips with the ecological realities affecting animals can be intellectually and emotionally overwhelming. We are, after all, contemporary witnesses to the early moments in a relatively rare occurrence: a mass extinction event that will permanently transform the nature and scope of animal life on the earth.[5] From a strictly scientific perspective, extinction refers to a state in which no members of a given species remain or when the remaining members of that species no longer exist in their native habitat (as, for example, when the few individuals of a species that remain alive are kept in captivity). The processes of speciation (the emergence of new species) and extinction are entirely normal aspects of evolution, with speciation typically outstripping the pace of extinction (hence the earth's characteristic biodiversity); and extinction of individual species tends to occur very slowly under typical evolutionary circumstances. This "background extinction rate" is difficult to determine with precision and must be inferred from fossil records and other indirect evidence.[6] What is undeniably clear from these records, though, is that the earth has seen previous periods where the extinction rate has spiked sharply, resulting in mass extinctions (typically defined as the earth's losing 75 percent of its species). There have been five previous mass extinctions, and many scientists suggest that the earth is currently in the middle of a sixth one, with the contemporary extinction rate being anywhere from 100 to 1,000 times the background rate.[7] What makes the present extinction rate especially noteworthy is that it is driven primarily by anthropogenic (human-caused) forces.[8] The chief anthropogenic drivers of the current extinction of animal species in particular include habitat destruction; pollution; overkilling and overharvesting; and climate change (an issue I will return to in the following section). All of these drivers are accelerating in their ecological impact, threatening even more extinction of animal species and loss of biodiversity in the coming decades and centuries.

This standard scientific framing of extinction in terms of species and biodiversity loss is useful for illuminating the scale of the present crisis, but it tends to gloss over the more profound meanings and consequences of extinction. As Eileen Crist notes, one of the great paradoxes of the present crisis is that it is occurring at the very time when we are beginning to gain a fuller appreciation for the richness and complexity of earthly life and animal minds.[9] Thanks to this increased awareness, many people are coming to recognize that a species' going extinct is not simply the loss of an abstract grouping but instead marks the end of an entire world, the annihilation of an intricate and irreplaceable way of life that has been built across millennia of evolutionary processes and relations.[10] Further, as Dominique Lestel observes, when a species goes extinct, not only is the natural world impoverished, but so is human life. He suggests that the ongoing loss of animal biodiversity represents a "psychological, intellectual, moral, and spiritual" impoverishment of humanity.[11] When a species goes extinct, myriad possibilities for cocreating new meanings, new relations, and new ways of being in the world go with it.

How might we begin to respond to the ongoing extinction event in a way that does justice to this deeper awareness of the stakes of such destruction? One key aspect of developing a fitting response is to approach extinction not just from the perspective of its implications for human life (say, as a loss of environmental "resources") but from the perspective of animals themselves. Along these lines, animal studies theorist Thom van Dooren urges us to think about extinction from what we might call the "underside" of the crisis—from the perspective of those animals facing annihilation.[12] For Van Dooren, this entails getting to know, in the deep ethological sense I have been describing, a particular species currently undergoing extinction as well as unique individual members of that species. To this end, Van Dooren has spent many years living with and carefully examining various birds (including albatrosses, vultures, whooping cranes, penguins, and crows), and he uses a detailed narrative and ethological approach to introduce readers into bird worlds and ways of life (which he calls "flight ways"). Immersed in these bird worlds, readers gain a richer sense for how birds live and die, and of what goes missing when their worlds and ways of life disappear. The task we face in the coming years is to multiply such experiments and narratives and bring them to bear on the transformation of our subjective, social, and environmental modes of existence.[13]

Ecological Degradation II: Climate Change

One of the main drivers of the extinction of animal species, as I have noted, is climate change. While readers are undoubtedly familiar with the general issue of climate change, we have (as with the problem of mass extinction) been encouraged to think about this phenomenon almost entirely from the perspective of how it affects contemporary human societies and future generations of human beings. The dominant culture gives very little thought to how climate change will affect animals themselves and the rest of the more than human world—beings and communities who have their own interests and their own stakes in the effects of ongoing ecological degradation. Let's try, then, to shift ourselves into a perspective from which this other set of concerns comes into focus. We should begin by reminding ourselves of some basic facts concerning climate change and then begin working our way toward a broader perspective on the issue.

Climate change denotes a range of phenomena that result from adding greenhouse gases (such as carbon dioxide and methane) to the atmosphere. It involves what is sometimes referred to as "global warming" (the accelerating warming of the earth's surface, atmosphere, and oceans), but it also encompasses a wide range of effects linked to global warming, including rising sea levels; changes in the location and intensity of precipitation; increases in the frequency and magnitude of natural disasters; and related phenomena. The rate at which human beings are emitting gases into the atmosphere has been climbing exponentially for decades, with atmospheric carbon dioxide concentration reaching 412 parts per million in 2020, a number that exceeds any level measured in ice-core records for some 800,000 years.[14] This rapid increase in carbon dioxide and other greenhouse gas emissions has led to a corresponding rise in global average land and ocean surface temperatures. From 1880 (the date when modern record keeping began) to the present, the global average temperature has risen 1.8°F, with the last decade being the hottest on record.[15]

In what ways do animals fit into the discussion about climate change? Perhaps the most evident relationship is the significant contribution that meat production makes to greenhouse gas emissions. The livestock industry creates emissions in a variety of ways, ranging from using fossil fuels in transporting animals and running farm equipment, to the release of methane by ruminant animals, to nitrous oxide emissions from tilled soils. Precisely how much livestock agriculture contributes to warming trends is a highly contentious and politicized issue. Scholars have offered estimates suggesting that livestock are responsible for anywhere from 10 to 50 percent of total global greenhouse gas emissions. The number most commonly used in the scientific literature is produced by the Food and Agriculture Organization of the United Nations, which puts the contribution of livestock production to total greenhouse gases at 14.5 percent.[16] (Beef and milk production makes up the bulk of these emissions, with pig, poultry, and egg production also making a significant contribution.) Even this fairly conservative estimate makes it clear that climate change cannot be adequately addressed without making significant changes to meat production and consumption; and the bulk of the scientific literature strongly supports the claim that vegan diets tend, on the whole, to be far more benign in climate impact than standard diets based on meat and dairy.[17]

Much of the pro-animal literature has limited the discussion of climate change to this dietary focus, but the problems surrounding climate change and animals run much deeper. As warming and other climate changes have accelerated over the past few decades, their effects on animals themselves have also become increasingly evident. No single story can be told about the impact of climate change on animals, insofar as the conditions for decline and flourishing on a rapidly changing planet vary considerably depending on the animals discussed and on their particular location. In general, though, climate change has tended to intensify the challenges faced by animals that are already in precarious situations. Many animals have been forced to change their ranges because of shifts in weather and food availability; and as species move to new habitats, they often encounter novel predators and diseases and sometimes crowd out other species already present. With global warming and the concomitant acidifying of oceans, entire trophic chains are being affected, with less food and fewer nutrients available for animals that often cannot find other sources of nutrition. Warming waters can also affect breeding rates among certain species, while loss of sea ice owing to warming waters stresses animals that use sea ice for cover and habitat. Although some animals will flourish in a rapidly warming and changing climate, many more will face serious difficulties and possible extinction.[18]

Note also that such changes in range of habitat, availability of food sources, and ecosystem composition associated with climate change are bringing human beings and animals into unprecedented contact. These points of contact constitute one of the most important aspects of the social ethology discussed in the previous chapter, since all such encounters have to be negotiated carefully in view of their potential for species' flourishing together as well as for the possibility of further degradation of wildlife, the introduction of zoonotic diseases, and so on. Climate change and a simultaneously expanding human population are producing a world wherein animals and humans are increasingly coming to live side by side in unaccustomed and crowded conditions. Whether human beings and animals can learn to live together respectfully under such conditions is one of the central social-environmental challenges we face in the coming decades.

Landed Animals

I began this chapter thinking along with Linda Hogan in order to elaborate the rudiments of an environmental ethology. As we have seen, Hogan's writings on animals recurrently direct our attention to the milieus where animals live, and they stress the need to attend to the ontological and ethical dimensions of those milieus. Perhaps the most evident domains for thinking about animal environments are the ecological ones we have been examining. We have considered how the ecological milieus where animals live are today being degraded along several lines, with the extinction and climate crises being among the most visible and urgent forms of ecological decline. Concern with the fate of animals in our time, I have suggested, necessarily includes attending to these ecological milieus and focusing on how rapidly degrading conditions affect animals themselves.

Yet Hogan's analysis of animal environments seeks to go beyond an ecological focus and encourages us to also consider the more expansive notion of *land*—a common and long-standing concern in Indigenous discourse and activism. For Hogan, relations with animals are always contextualized and nested within milieus comprising a variety of what we might call "third parties" that include ecological systems but exceed the scope of this set of relations. Kim TallBear (Sisseton-Wahpeton Oyate), another influential thinker in contemporary Indigenous studies and animal studies, provides us with a useful way of understanding the stakes of this more expansive view of environment as land.[19] While TallBear is a recognized participant in, and appreciates, the general trends in animal and environmental studies that have led to an increased focus on animals and the more than human world, she questions whether recognizing the social and ecological dimensions of the animal question alone is sufficient. In raising this concern, TallBear does not in any way minimize the importance of focusing on more than human social and biological relations or interspecies communities (say, in favor of an exclusively human-centered orientation); indeed, she takes pains to stress that Indigenous traditions have long emphasized such relations and that animal studies and the environmental humanities owe a considerable debt to these traditions for sustaining non-anthropocentric perspectives.[20] Rather, TallBear's concern is that most non-anthropocentric trends in mainstream theory simply do not go far enough in thinking about the extent of relations that constitute the communities and milieus where we dwell. In particular, she points to a questionable vitalist and biocentric streak within mainstream non-anthropocentric frameworks, a tendency that limits our "attention to beings that 'live,' e.g., dogs, bears, mushrooms, microorganisms. We speak of animal studies after all."[21] TallBear notes that for many Indigenous peoples, the nonhuman others with whom relations are formed extend well beyond living beings; hence, heightened focus on living individuals might come at the expense of paying equal attention to relations in other registers (for example, relations to inorganic beings and systems, to heritages and ancestors, to the processes of decay, destruction, disintegration, and so on).

In this vein, many Indigenous traditions consider land—which includes a much broader set of relations—to constitute the center of community; in fact, it is often understood that the meanings and values that bind a community together inhere in the land itself.[22] Meaning and value here are not human constructs imposed on reality but are discovered and co-constituted in relation to the land itself. According to Glen Sean Coulthard (Yellowknives Dene), it is this normative dimension of land that often undergirds the remarkable resistance many Indigenous peoples display to commodification of their territories.[23] Coulthard notes that it is a profound misunderstanding to equate Indigenous resistance to the commodification of land with the protection of private property, for land in this context is less some*thing* that can be delimited and more a dynamic and expansive set of relations to which one is responsible. Land is thus not an object to be owned but a complex field of relations and processes by which a people is formed and sustained. Land includes what is typically understood by that term in colloquial American English (the material earth surface in a particular area), but also extends to "people and animals, rocks and trees, lakes and rivers, and so on."[24] We should be careful here, though, not to be misled into thinking we can discover what land means by making a comprehensive list of all such human and more than human beings in a given region, for the key ontological category in considering this Indigenous conception of land is *relation*. To attend to the land is to attend to the relations that constitute particular places—which is to say, to the complex web of individual and extra-individual forces, systems, and assemblages that constitute the world of a given people or community.

Coulthard explains that, within his own tradition, this relational notion of land carries three important corollaries or consequences. First, the gesture of making land central to life decenters human beings, rendering them no more (and no less) significant than any other element of the land. Second, there is no sharp ontological rupture between human and more than human beings in this conception of landed existence. The sorts of capacities we might typically associate exclusively with human beings (Coulthard refers to "spirit or agency"[25]) are understood to permeate the land in varying ways and to varying degrees. Third, ethical obligations based on shared agency or subjectivity cannot be limited to the interhuman realm, insofar as such characteristics and relational capacities are shared. As Coulthard notes, given the reciprocal notion of ethics at work here, obligations are understood to run in both directions; if human beings fulfill their obligations, they can expect that the land and other relations will do the same.[26]

We can see, then, that this non-anthropocentric conception of land is onto-ethological and onto-ethical in nature. Land is understood to comprise a set of human and more than human interrelations that expand the standard notion of what constitutes the lived environment. In this sense, land serves as a critical ontological category that emphasizes not just the primacy of relations but their dynamic and expansive nature. Shared milieus comprise relations of myriad kinds, interconnections that create unanticipated possibilities for living and dying in common. At the same time, this ontological vision is linked with an ethical register, in which relations are not only acknowledged in their reality but affirmed in their goodness. As Coulthard suggests, the reciprocal relations that constitute land provide us with a vision for how we might live "our lives in relation to one another and the natural world in nondominating and nonexploitative" ways.[27] From this perspective, norms reside in the land itself, in the very form that landed relations take and in the ways those relations engender and sustain cycles of life and death. Coulthard uses the term grounded normativity to refer to this ethical register of relation and suggests that the reciprocal relations that sustain the land across time might come to "inform and structure our ethical engagements with the world and our relationships with human and nonhuman others over time."[28]

For readers familiar with the history of environmental thought, this concept of land might seem to smack of the place essentialism that plagues certain environmental ethical frameworks (such as ecocentrism and bioregionalism), in which a given place (say, a particular ecosystem or bioregion) is understood to have stable, clearly defined boundaries and where identity and norms can be straightforwardly derived from that territory. Environmental theorist Ursula Heise has justly criticized this sort of environmentalism for its localism and provincialism and has suggested that environmentalists should instead "foster an understanding of how a wide variety of both natural and cultural places and processes are connected and shape each other around the world, and how human impact affects and changes this connectedness."[29] While perhaps not apparent at first glance, the Indigenous concept of land we have been examining with Hogan, TallBear, and Coulthard already includes the richer and broader sense of environmental milieus that Heise propounds. Although the term land as used by these Indigenous theorists is indeed intended to include reference to *these* relations (in this particular place) rather than *those* relations (in another place), the lines delimiting relation are by no means understood to seal one place off from another but rather are seen as flowing outward to connect lands and, in turn, to link these various places to the planet as a whole.[30]

Non-Indigenous individuals might wonder how to respond to these land-based discourses and practices. Some readers might also wonder more specifically about the relevance of the notion of land to pro-animal discourse and activism of the sort examined in the previous pages. Ecologist Robin Kimmerer (Potawatomi) provides a helpful way of approaching these questions in the context of her discussion of a lawsuit by the Onondaga to reclaim title to their ancestral lands in Upstate New York.[31] The heart of that territory is Onondaga Lake, which in 1994 was listed as a Superfund site by the Environmental Protection Agency and bears the scars of decades of irresponsible commercial activity. For the Onondaga, as Kimmerer explains, this lawsuit is an effort to regain the land title "not for possession, not to exclude, but for the right to participate in the well-being of the land . . . , for freedom to exercise their responsibility to the land."[32] While becoming responsible for the land is, to be sure, a laudable goal, settlers might be unsure about their role in such a project. Are there positive roles for settlers here? Or does the movement for repatriation of Indigenous lands entail settler exodus and resettlement? While answers to such questions will vary based on context and the land claims under consideration, Kimmerer suggests that in the case of the Onondaga, repatriation is not a matter of forcing people (settlers or otherwise) from their current residences but is rather an invitation to become jointly responsible for the land:

> I think that the land rights action is an invitation for the people of this watershed to engage in becoming Indigenous to place. No newcomer can ever match the Onondaga's identity with these hills, but what does it mean for an immigrant culture to start thinking like a native one? Not to appropriate the culture of Indigenous people, not to take what is theirs, but to throw off the mindset of the frontier. . . . Being Indigenous to place means to live as if we'll be here for the long haul, to take care of the land as if our lives, both spiritual and material, depended on it. Because they do.[33]

Following Kimmerer's suggestion, we might consider how this Indigenous approach to land could inform the ongoing development of animal studies. The main theoretical frameworks that tend to dominate in the field—animal rights, animal liberation, total liberation, abolitionism, posthumanism, companion species approaches, and so on—have generally paid scant attention either to the history of settler colonialism or to how projects for animal justice relate to the ecology and the land more broadly. How might animal studies be transformed if it took on board this sort of vision of being responsible to the land? What kind of questions would come to the fore? How might current campaigns and struggles at the heart of the animal rights movement be reassessed and reconfigured? I am not trying to determine how this critical encounter should unfold. Rather, I want to emphasize that Indigenous struggles over land can and should serve as an opportunity for the field of animal studies to consider more fully its connections with settler colonialism and how practicing a more respectful relationship with the land might complicate but also expand and enrich human-animal relations.

Although not exclusively focused on land, a body of work currently emerging from scholars and activists across the theoretical spectrum is dedicated to exploring some of these themes. For example, Billy-Ray Belcourt (Driftpile Cree Nation) has highlighted fundamental interlinkages between settler colonialism and anthropocentric attitudes toward animals and argues that thinking through both forms of oppression together is mutually illuminating and politically essential.[34] This position has gained added support from the "triadic approach" advanced by Fiona Probyn-Rapsey and Lynette Russell (Wotjabaluk), which calls for more consistent efforts to bring considerations of colonization, indigeneity, and animality into dialogue.[35] From a legal studies perspective, Maneesha Deckha has offered a compelling case for developing protections for animals based on anticolonial and feminist theories of embodiment and vulnerability rather than on traditional Western notions of legal personhood.[36] In recent years, we have also seen high-profile conferences and special issues of journals focused on the intersection of colonialism and animals, featuring a range of Indigenous and animal studies scholars.[37] In terms of activism, collectives such as the Farm Sanctuary are working to educate the public on the connections between animal exploitation and the historical legacy of colonization and ongoing forms of settler colonialism, while new coalitions are currently being forged among direct-action animal rights activists and anticolonial activists.[38] In important ways, this recent body of work can be seen as an extension of long-standing trends among pro-animal feminists and ecofeminists, many of whom have sought to develop more expansive notions and practices of intersectionality beyond traditional sociological categories.[39] Among the most important figures in this vein is Val Plumwood, who some three decades ago emphasized the overlapping logics of colonialism and the oppression of women, ecology, and animals, and who continued to explore these connections with great insight up until her final writings just before her death.[40]

Of course we should not expect an easy commensurability between the pro-animal movement and Indigenous critics of settler colonialism and defenders of land. While there is considerable overlap between these movements in terms of their respective critiques of anthropocentrism, there remain incommensurabilities in their goals and worldviews, as well as understandable concerns about the abuse and misuse by settlers of Indigenous ways of knowing and forms of life.[41] Such differences, though, need not prevent political work and coalitions among these movements; in fact, attending carefully to these differences is the very condition for enacting genuine politics. Moreover, to adequately address the social and ecological crises that plague human and more than human life, we have little choice but to continue working humbly and patiently at this intersection.

What Is Environmental Ethology?

Keeping in mind the foregoing discussion of ecological issues and Indigenous ideas about land, we are now well positioned to think more carefully about the nature and stakes of environmental ethology. Defined in the most basic terms, environmental ethology involves attending carefully to the abodes where animals dwell and we dwell with them. But *abode* and *dwelling* carry several senses that are relevant to this dimension of human-animal relations. In examining the various senses of these terms, it is essential to underscore that the attention given to animal environments is driven by a fact that is very basic but easily overlooked: if we are concerned about the flourishing of animals today and in the future, we must take into account their lived environments. Animal liberation means very little if animals are liberated into a world that is so degraded they cannot flourish in it. Thus, environmental ethology begins from the premise that pro-animal discourse and activism must from the outset bear firmly in mind an environmental sensibility and disposition.

Precisely how we ought to think about and weigh relations between animals and their environments has long been a contentious issue in animal ethics and animal studies. So we should first explore some of the obstacles that have prevented the two registers from being more deeply aligned. One such obstacle is the tendency within pro-animal discourse toward *methodological individualism*, which emphasizes the rights and welfare of individual animals and tends to leave their environments in the background. It is typically assumed within this discourse that, once our attention leaves the realm of sentient animals, we will encounter no beings who have subjectivity; hence our normative concerns can justifiably be limited to human-animal interactions and to the realm of beings who exhibit some individuality. As I have previously noted, I believe the pro-animal emphasis on individual subjectivity is something worth preserving, especially in view of an established order that refuses to grant even basic standing to the billions of animals it slaughters, tortures, and sacrifices yearly. That said, we should not allow the normative structures of the status quo—structures premised on the dogma that the only beings who truly count are those who can occupy a certain subject position—to mislead us into assuming that *our* normative concerns are coterminous with the community of those deemed to be full and proper subjects.

Subjectivity exists in varying degrees throughout human and more than human environments, and we can learn to attend to those varying forms in different ways. Likewise, if we find places where subjectivity (in the strong sense of the term) tends to shade off and other forms of existence start to emerge, there is no need to cut off our ethical attention and concern; we can instead learn to be attentive and show care in ways suitable to those other types of beings, systems, and relations. Indeed, a range of environmental and more than human ethical frameworks seek to do just this, and there is no need to assume that there is an intractable antagonism between concern for animals and attentiveness to nonanimal beings and relations.

Another difficulty that pro-animal theorists have encountered in attempting to think about the environmental dimensions of animal issues is what could be called *intentionalism*, or the tendency to emphasize the importance of people's intentions when making decisions. In adjusting our relations with animals, our first (and quite reasonable) impulse is often to think about the direct harms we as individuals are causing through our intentional actions and then trying to change those behaviors. Thus we tend to focus on what we deliberately decide to eat and wear, what medicines or cosmetics we purchase, and so on, and aim to make changes along those lines. But when considering broader questions of responsibility for ecological degradation (and not just direct harms to animals), the question of how to proceed is more complex, and one's individual intentions are not always the most salient concern. The harms at issue in an ecological register are often the result of indirect and unintentional acts; the sources of the harm are sometimes difficult to pinpoint; the beings or systems suffering harm can be challenging to delimit and are often transindividual; and the specific contributions we make as individuals to the problems at hand are sometimes impossible to determine.

Thus, when considering ecological issues, the ethical and ontological frameworks that are most common in pro-animal discourse often run up against intractable limits, requiring a shift toward systemic and structural analysis that can seem foreign to a discourse that has its origins in liberal humanist commitments (where individuality and intentionality play a central role). It would be reassuring to believe we could adjust our environmental relations through political liberalism and individual responsibility at the level of consumer purchases and life decisions, but the problems in this domain require a very different way of proceeding, one this book has been striving toward.

To pursue this environmental ethological line of thought further, let's return to the work of Joe Hutto, featured in chapter 1. Throughout his reflections, Hutto emphasizes the way his ethology of the mule deer opens him to specifically environmental considerations and insights. In following the deer and learning to move increasingly within the herd, he comes to understand more fully just how deeply the deer are embedded in their specific ecological niche. Ecological relations are so fundamental to the deer's way of life, in fact, that Hutto starts to see the deer as indiscernible from those relations. In the following passage, he offers a powerful description of the vision he attains once he finds himself fully accepted by the herd and comes to appreciate how fundamental their milieu is to their being:

> Stunned, my improbable and privileged perspective became crystal-clear, as with new eyes I began to see the near-impossible beauty of this perfect part of a perfect landscape. Mule deer, sage brush, snow, and mountain became indistinguishable. Where did mule deer end and sage brush begin?

> This singular vision of a creature so perfectly interwoven into the ecology instantly transformed the way in which I perceived this remarkable animal, but also forever changed my understanding of its significance as an indivisible component within the landscape.[42]

To affirm and be attentive to an individual mule deer entails, from this enlarged perspective, simultaneously appreciating this particular deer's relation to its herd and the broader ecology; in other words, one cannot be fully responsible to individual mule deer without appreciating that they are irreducibly nested within various relations. Thus there can be no absolute priority given to one perspective (say, the individual mule deer or the herd as a whole) over another (say, the broader ecological systems deer live and participate in). Rather, to practice environmental ethology in Hutto's manner, we must constantly bear multiple registers in mind. To remain with the mule deer and live in and among them, Hutto must practice a perspectival and relational pluralism, attending to specific personalities and histories of individual deer, to herd dynamics, and to salient environmental factors.

Hutto's heartfelt plea to readers to protect and preserve the ecology of the mule deer he came to love and cherish might make us recall that these deer are not alone in their fight against extinction and degraded habitats; as I have noted in this chapter, myriad other species are also on the brink of disappearing, forming part of a larger pattern of mass extinction, defaunation, loss of biodiversity, and general environmental degradation that affect animal life as a whole, whether they are domesticated, liminal, or wild. Consequently, to sustain extant forms of human-animal relations and to maximize the potential of those relations for individual and collective transformation and flourishing, the mental and social ethologies we have discussed must straightaway become linked—conceptually, practically, and ethically—with an environmental ethological disposition.

When undertaken with a deep ethological sensibility, then, the reconfiguration of human-animal relations opens onto the realm of environmental ethology and faces the challenge of bringing about a fundamental reconstitution of our collective relations with the ecologies we find ourselves in. I have highlighted climate change and mass extinction in this chapter, but expanding animal studies into an environmental ethology would mean paying greater attention to and establishing more direct linkages with movements for transforming systems of transport away from reliance on personal vehicles and toward more sustainable and less violent forms of mobility; developing alternative and less harmful modes of energy use; and reforming urban infrastructure to minimize disruption of surrounding ecologies, to name some of the more prominent and pressing issues. In addition to the more familiar struggles against animal agriculture, hunting, and experimentation that have formed the core of the pro-animal movement, all these environmental issues should be seen as being of immediate and direct concern, for they highlight the practices and institutions that degrade the ecological milieus of animals and the life-forms and systems within which animals dwell with us and other planetary kin.

As we have seen, the practice of environmental ethology also entails going beyond the ecology to include a broader sense of the environment as a specific place and dwelling site, constituted by innumerable relations. The ethological experiments I have discussed in these pages obviously unfold in a particular milieu. Hutto's relations with the mule deer are specific to the mountain range in Wyoming where he and they reside. My and my family's relations with Patches and his murder emerge from the peninsular range of Southern California and would not be exactly replicated elsewhere. Hogan's numerous animal relationships are unique to Idledale and the particular milieus and networks within which those relationships have been constituted. In all three cases, the lands where these relations have been formed are colonized lands, a point to which Hogan as an Indigenous thinker is particularly sensitive and to which pro-animal discourse and activism have been rather less attentive. In extending the notion of milieu beyond a strictly ecological perspective, elaboration of an environmental ethology endeavors to avoid replicating the all too common tendency to overlook both the ongoing wounding of settler colonialism and the forms of life that settler colonialism has interrupted.

By attending to the land where human-animal and animal-animal relations unfold, environmental ethology helps us engage in two complex but necessary undertakings in this vein: first, developing a more expansive ethical sensibility; second, coming to grips with the colonial wound. Taking into account the concept of grounded normativity and the expansive notion of relation promoted by Hogan, TallBear, Coulthard, and related Indigenous thinkers, animal studies might come to see itself as one discourse among several related approaches that challenge anthropocentrism and its norms and institutions. To insist, as animal studies generally does, on the ethical and relational importance of animals, their singularity and their radiant lives (to use Hogan's term), need not prevent us from thinking about the ways value and importance circulate alongside and beyond sentient beings. In this sense, animal studies might be understood as a central but only partial exploration of a zone of inexhaustibly complex and dynamic relations that exceed the scope of the human. In adopting a disposition in which ethical consideration is taken to be inherently open-ended, dynamic, and experimental, animal studies can find a point of critical intersection with Indigenous notions of land and dwelling that have long sought this sort of maximally expansive notion of relation.

Second, by addressing the ongoing consequences and harms caused by settler colonialism, pro-animal discourse and activism can correct some of the oversights of previous generations of theorists and activists who tended to think of justice for animals as divorced from such concerns. Achieving a more just and more respectful world for animals need not be viewed as a zero-sum game; this project can take its place alongside a variety of other movements with related aims and enter into productive relations with them. With respect to resurgent Indigenous politics in particular, pro-animal discourse stands to learn a great deal, for these transformational struggles are among the very few contemporary political struggles that are grounded in non-anthropocentric worldviews and normative frameworks. As I have noted, the ideal of developing just relations with animals has a longstanding and prominent place in many Indigenous traditions; and even if these ideals do not always perfectly align with the aims of animal liberation or animal rights, the points of overlap and divergence can produce novel theoretical and practical insights.

Anticolonial Vegan Ethologies

The work of Mi'kmaq scholar and activist Margaret Robinson offers a particularly illuminating example of the perspective that can be achieved by bringing contemporary pro-animal discourse and activism into dialogue with struggles against settler colonialism.[43] Robinson's work establishes this dialogue around one of the most contentious issues involving animals: veganism. Although Robinson is herself a vegan, she does not posit a simple commensurability between this way of eating and Indigenous foodways; rather, she develops a complex, transversal relation between the two traditions that brings them into dialogue and allows them to challenge and inform each other.

Robinson realizes there are cultural and ideological barriers that might seem to render any alignment between veganism and Indigenous cultures unlikely. There is, first, the apparent "whiteness" of veganism, insofar as it is commonly portrayed as a diet promoted and practiced primarily by white people. Further, there is the apparent class privilege attached to veganism, inasmuch as only those who live where they can obtain vegan foods and meat/dairy substitutes (which can be prohibitively expensive) can practice this way of eating. Robinson, of course, acknowledges the importance of race and class in the constitution of foodways, but she also contests the notion that veganism is reducible to whiteness and class privilege. Readers familiar with the history and contemporary practice of veganism will know that the stereotype (that veganism is practiced primarily by economically privileged white people) tends to gloss over a much more complex reality. Vegan foodways, as well as vegetarian diets and other forms of eating that take animal well-being into account, are hardly a contemporary invention of white people. They have roots in both Western and non-Western cultures that reach back thousands of years; moreover, the sorts of foods most vegetarian and vegan people eat are, in fact, much closer to the foods eaten by poorer people across the globe than to the meat-heavy diets of many Westerners. Although we cannot deny that there is a segment of modern-day vegans who are white and class-privileged and who eat extremely expensive foods with large carbon footprints, this subset cannot fairly be made to represent global vegan and vegetarian populations.

Robinson would also have us note that the diets of many contemporary Indigenous peoples are in fact already colonized and are often not particularly healthy or sustainable. Citing Kim Travers's research on Mi'kmaq diets, Robinson notes that the main causes of poor diet among her people are "low income, lack of access to transportation, and reserve land unsuitable for agriculture, fishing, or hunting" (190). This situation leaves Mi'kmaq people living on their reserve land "eating highly processed protein such as peanut butter, wieners or bologna" (190)—hardly the fare one would wish to consider "traditional" or "authentic." So the issue here is not substituting a diet exclusive to white, privileged Westerners for a healthy, traditional diet but inquiring into what it means to eat well today in a world that has been ravaged and utterly transformed by colonialism, capitalism, and ecological degradation.[44]

One possible solution for Indigenous peoples in addressing these challenges would be to return to traditional foodways, leaving behind colonized diets as well as vegan alternatives (where the latter might also be seen as a colonial imposition). There are in fact proponents of this strategy, many of them linked to the ongoing struggle to maintain fishing and hunting rights on traditional lands.[45] In response to this approach, we must acknowledge that the traditional diets of many Indigenous cultures did give meat a prominent place. Here too, then, there seems to be a fundamental incommensurability between veganism and Indigenous ways of life. Robinson acknowledges this possible incompatibility and is by no means insensitive to the politics of identity, land rights, and tribal sovereignty that are often linked to recuperation of traditional foodways. But she also seeks to reframe the discussion by noting that the contemporary context of the discussion over foodways has changed considerably from the circumstances in which traditional eating practices were formed. In this vein, remaining faithful to one's heritage and relations is, for Robinson, not simply a matter of reinstituting traditional practices but one of discerning how best to enact the tradition's underlying principles and vision for what it means to live well in ever-changing contexts. From this perspective, she would have us—both Indigenous and non-Indigenous peoples—ask whether eating meat *today* (and especially the sort of meat that is now widely available) aligns us more closely with the logic and institutions of colonialism or with decolonial resistance.

In answering this question, Robinson returns to the foundational myths of her Mi'kmaq tradition. Much like Hogan, Robinson distills from traditional narratives an ontological as well as a normative view of human-animal relations. She suggests that the vast majority of Mi'kmaq legends provide a view of "humanity and animal life as . . . [existing] on a continuum, spiritually and physically" (191) rather than being separated by an insuperable divide. Robinson explains that the archetype of human existence in Mi'kmaq legends is Glooscap, a being who is directly constituted by earthly materials and forces, has ancestors who are formed from and represent various parts of the natural world, and has extensive relations with the more than human world. Although killing animals is permitted in traditional Mi'kmaq stories, Robinson stresses that killing with impunity is generally disallowed. The overarching vision regarding relations with animals, she suggests, is one of human dependence, not dominion; and when themes of dominion or mastery do appear, the use of animals for human ends is restricted by necessity. Broadly speaking, then, animals are seen in these narratives as beings who have an independent existence and who, according to a reciprocal economy of giving and respect, allow human beings to use them only under certain conditions and only with their consent. The elaborate rituals that surround killing and eating animals also suggest that their death is ultimately regrettable and unfortunate and hence to be limited strictly to the fulfillment of genuine need.

On Robinson's reading and interpretation, these traditional myths are intended not to tell Mi'kmaq people exactly how to live but rather to provide them with a persisting set of overarching values and a general orientation in the world. A narrative-based worldview of this sort, then, must always be contextualized and enacted in a dynamic cultural and natural setting. Thus traditional stories require exegesis and thoughtful application rather than uncritical appropriation or imitation. When read and interpreted in view of the contemporary context of Indigenous struggles and global animal and ecological issues, Robinson suggests that traditional Mi'kmaq stories might give rise to some surprising counter-readings. So, rather than seeing veganism as inauthentic for the Mi'kmaq and other Indigenous peoples, she argues that the themes of animal personhood and self-determination, and of human regret over death, as they appear in these narratives suggest that "choosing not to ask for [animals'] sacrifice is a legitimately Aboriginal option" (193). By contrast, strictly identifying Indigenous peoples with traditional practices of hunting and meat eating, Robinson maintains, locks them into a static, precolonial mode of existence that prevents them from adapting and responding to contemporary conditions. If living well in view of and with regard to the land is the overriding ethical principle of her tradition, it is essential to recall that there is "more to [Mi'kmaq] culture and to our relationship with the land, particularly as women, than hunting and killing animals" (193).

Thus veganism, when approached from the perspective Robinson articulates, can become another means of fulfilling ethical obligations and of embodying "traditional values in new rituals" (194). Viewed as a ritualized, spiritual practice, veganism serves as a means of recalling "our connection with other animals" (193) and prefiguring another way of life (to recall a gesture we encountered in Hogan's work). Thus, rather than characterizing indigeneity as an unyielding commitment to maintaining static traditions, Robinson suggests that Indigenous cultures should be understood as "living traditions, responsive to changing social and environmental circumstances" and ways of life that continually seek more finely tuned ways of responding to the needs of animal kin and the "needs of the land itself" (194).

While vegan readers might be tempted to see Robinson's argument as a straightforward endorsement of mainstream vegan discourse and practice, here it is important to proceed with some caution.[46] In particular, given the focus of this chapter, it is essential to note the role that ecology and land play in Robinson's veganism. While her version of veganism is primarily a way of doing justice to animal relations, those responsibilities are already embedded in a wider set of responsibility to *all* relations, including ecology, the land, and other peoples. Thus, for Robinson veganism is both part of a larger ethical vision and a means of recalling readers to that broader set of relations. On this approach, veganism and respect for animals cannot be reduced to a simple change in consumer preferences but should instead be seen as leading to a thoroughgoing subjective transformation brought about by ritual practices and a sustained commitment to becoming different kinds of selves. Likewise, it asks us to adopt a different social ontology, one that is less concerned with showing how animals are "like us" and more interested in exploring how human beings are latecomers to an ongoing set of relations where we are plain members. Ultimately, then, the sort of veganism Robinson envisions is not recommended as a stand-alone practice for bourgeois white people in advanced capitalist societies but as a central component in the decolonial project of enacting respectful relationships with animals, the land, and the myriad others we are connected with. An environmental ethology that takes up this call would thus be thoroughly ecological in orientation but would also learn to place the animal question inside the history and ongoing present of settler colonialism and the reconstitution of respectful relations with the land.

CONCLUSION

A Worthwhile Life (and Death) with Animals

In the previous chapters I have presented the three ethologies as serving two main ends. I have sought first of all to demonstrate that animal studies and other forms of contemporary pro-animal discourse and activism might benefit from adopting a more comprehensive, multipronged ethological framework, one that aims to transform the registers of self, society, and environment. Second, I have sought to show how these various registers can be brought to bear on one another to highlight possibilities for change that might otherwise go unnoticed. I have referred to such interactive and intersecting relations between registers as *transversal*, a key term in the work of Félix Guattari, whose writings have inspired the basic framework of this project. Bringing the different ethological registers into a transversal relation has allowed them to work through and against each other, to mutually inform one another, and to open up fresh potentials for thought and practice. Along the way, I have also suggested that the task of reconstituting human-animal relations at the subjective, social, and environmental levels can help to broaden the political vision of pro-animal discourse and activism and give it the impetus to connect more deeply with a variety of related movements for social and environmental transformation. And where work of this sort is already under way, I hope the rubrics and concepts developed here will help give form and coherence to those efforts and make their approach more compelling to a broader audience.

Beyond these ends, I suggest in this conclusion that the practice of the three ethologies also constitutes a meaningful and worthwhile way of life. In brief, I suggest that forming better human-animal relations is an instance of the good life, or at least *a* good life. Further, I touch on how this way of life can reframe and reanimate central philosophical concerns in ethics, ontology, and aesthetics as well as justify the pursuit of an animal-philosophical mode of existence. To provide a full explanation of these themes would require its own book, but here I want to at least to give some indication of how the practice of the three ethologies might constitute a life well lived.

First, though, we should ask: Whence this concern for living a worthwhile life? Although I have used a variety of discourses and traditions in these pages, at heart the book's concerns are philosophical in nature; and philosophy, at least as I understand it, takes living a worthwhile life as its primary concern. If you have been exposed to the study of philosophy in a university, this conception of philosophy might stand somewhat at odds with your experience. In universities, philosophy is often treated as one discipline among others, with the specific task of transmitting to students a certain set of skills (for example, the tools of critical thinking and conceptual analysis) and disciplinary content (for example, material from the history of philosophy, current debates in metaphysics and epistemology, and so on). Traditionally, however, philosophers have had a very different understanding of what is at stake in doing philosophy. Although the practice of philosophy has often included the study of logic, epistemology, metaphysics, and other key branches of inquiry, in the classical era philosophy was understood to be something much broader and much grander than a method for achieving intellectual competence in these fields. For the ancients and for many subsequent philosophers, practicing philosophy was understood to involve converting to an entirely new way of life. As scholar Pierre Hadot (who is largely responsible for recalling for us today this historical vision of philosophical practice) notes, philosophy as a way of life involved a complete transformation in one's subjectivity and the adoption of an entirely different way of being in the world.[1] In some cases, as with the ancient Cynics discussed in chapter 1, such differences were stark and visible and involved strikingly unusual ways of dressing, eating, and being in public; in other cases, as with the Stoics and Epicureans, the changes were subtler but still involved practices and habits markedly different from those of non-philosophers. No matter what form these changes might have taken or what particular ideals a school might have promoted, adopting a philosophical way of life almost always entailed trying to redirect one's day-to-day existence onto the path of a true and worthwhile life (with the assumption being, of course, that the way of life on offer in the dominant culture was ultimately devoid of genuine value and significance).

Syn-theōria, or Seeing-With

The three ethologies sketched in the preceding chapters can be fruitfully understood as a contemporary version of this classical philosophical project of developing a worthwhile way of life. The deep ethological practices I have described do not, of course, seek to replicate the specific practices or worldviews of classical schools of philosophy; the intellectual and cultural conditions under which those schools emerged are far removed from our contemporary subjective, social, and environmental contexts. But we can always endeavor to re-create a meaningful way of life in our own age and context, inspired by previous attempts to do so.

Perhaps the most salient connection between the practice of the three ethologies and the classical vision of philosophy as a way of life lies in the practical ideal animating the present project—an ideal I refer to as *syn-theōria*, or seeing-with others. Throughout the previous chapters, I have emphasized the ways appreciating animals' perspectives and learning to see-with animals is essential to transforming subjective, social, and environmental relations; moreover, as I have noted at multiple points, there is an affirmative *passion* for such alternative modes of beholding the world with and alongside animals that motivates deep, philosophical ethologies. The general passion for beholding the world and seeing it with fresh eyes—an activity the ancients called *theōria*—has a venerable pedigree. For philosophers like Pythagoras, Plato, and Aristotle, *theōria* named the supreme activity of philosophical life, such that a worthwhile existence was to a significant degree one of learning to see in the deeper sense of having a philosophical vision of things. This deeper mode of vision was contrasted with the largely illusory and misguided worldview belonging to common sense and the dominant culture. Learning to genuinely see, to theorize (in the sense of practicing *theōria*), involved setting aside the established habits of seeing and thinking in which one was immersed from birth and striving to develop new habits and dispositions.

To understand the significance of reframing the traditional practice of *theōria* as involving our fellow animals (as I am seeking to do here), let us examine in more detail how this term was originally used and how our usage builds on and departs from this tradition. Originally, the group of terms associated with *theōria* (which includes cognates such as *theōros*, *theōreō*, and so on) referred to the act of carefully observing and looking at something, especially in the sense of being a viewer or a spectator at some event.[2] Some of the oldest uses of *theōria* refer to a person's being sent as a political delegate to visit another community or to an event such as the Olympic Games in order to observe the customs and habits of a people and to report their observations back home. In taking over this cluster of terms and repurposing them for themselves, ancient philosophers sought to show how careful observation of fascinating spectacles is akin to the philosophical study and observation of the world in all its wonderful and astonishing forms. *Theōria* for philosophers came to name the activity specific to those who have left behind the flattened, dull, and well-worn perspectives of the established order and common sense in order to strike out and see the world anew and truly (symbolized, for example, by the familiar story of the cave in book 6 of Plato's *Republic*).

Perhaps the most elaborate and influential philosophical account of *theōria* we have from the ancients is found in Aristotle. He argues that the only intrinsically worthwhile activity is *theōria*; all other activities, by contrast, have ends beyond themselves. Moreover, he maintains that *theōria* is the sole practice wherein genuine human flourishing ultimately comes to fruition.[3] But what exactly is *theōria* for Aristotle? It is, of course, a form of seeing, but it specifically involves learning to catch sight of things in a more profound way than can be gained by simply looking at them. To practice the sort of seeing involved in *theōria* is to perceive the deep and persistent structures of being, the very nature and causes of things. *Theōria*, then, is the act of comprehending how things hang together in a meaningful and coherent way. Contemplating and beholding things in this manner, according to Aristotle, not only is intrinsically rewarding, but also requires nothing but one's own intellectual faculties. In this sense, *theōria* renders human beings akin to the gods: self-sufficient beings who live solely to contemplate the wondrous nature of things. Aristotle is well aware that *actual* human beings will never fully achieve or sustain such a state; but to the degree that we do in fact practice *theōria*, we thereby participate in the only semidivine activity to which human beings have access.

While Aristotle's conception of *theōria* is useful for illuminating some of the key aspects of what is at stake in our correlative notion of *syn-theōria*, there are clear differences between the two approaches. The most noteworthy difference is that the ideal of *syn-theōria* animating the three ethologies presumes that *animals themselves theorize*—which is to say that animals behold the world in modes specific to their subjectivities and communities, and that they do so in ways that are worth trying to understand, respect, and even inhabit. Aristotle, by contrast, flatly denies that animals are capable of *theōria* or its attendant happiness. He maintains that animals are "entirely deprived" (*esterēmena teleiōs*) of and "share not at all" (*oudame koinōnei*) in contemplative activity, and hence they never experience the blessedness and happiness human beings are capable of.[4] For readers who know something about Aristotle's broader ethical philosophy, such a conclusion should come as no surprise, since he maintains that intellect (*nous*) distinguishes human beings from other animals and that *theōria* is the active fulfillment or proper enactment (*energeia*) of this distinctive characteristic. The ontological standpoint of indistinction that I have adopted from the outset of this work starts from a very different set of premises—that there are no such radical breaks between human beings and animals and that the traits we take to be distinctive of human beings are widely shared with animals and other registers of the more than human world. Thus, even if animals do not theorize in precisely the ways Aristotle defines the term, we are unable to follow him in concluding that they do not see and behold in any way or that they have no share in the sort of flourishing and blessedness he reserves for human beings.[5]

Another key difference in these respective approaches to the practice of *theōria* concerns the relational nature of seeing-with. One of the things that recommends *theōria* to Aristotle as the highest ideal is, as I just noted, that it can be practiced without relying on others, thereby allowing for self-sufficiency (*autarkeia*). *Syn-theōria*, by contrast, is a communal form of seeing, one possible *only* through world-traveling and by learning to see with and through others.[6] This makes any particular practice of *syn-theōria* susceptible to being interrupted or even broken entirely, a possibility that mirrors the vulnerability and mortality of its earth-bound practitioners. Thus, even as Aristotle emphasizes the ways in which the stability and self-sufficiency of *theōria* link the contemplative philosopher with the divine, *syn-theōria* stresses our partial, embodied, and finite natures and affirms our inexorable immersion in contingent earthly relations.

I should underscore as well that *syn-theōria* is distinct in what it seeks to behold. As I stated earlier, for Aristotle *theōria* is directed at perceiving the persistent principles and causes that structure reality and give it its underlying grain and direction. *Syn-theōria*, as conceived here, does not reject a priori the search for deep patterns, but neither does it restrict itself in this way. It begins from the premise that the richest picture of reality is gained by multiplying perspectives, learning to inhabit different worlds, and exploring the possibilities for finite life that are found there. Thus, in adopting a syn-theoretical approach no initial decision is made in favor of the One (relatively stable patterns) or the Many (radical change and novelty); instead, this approach proceeds from a commitment to exploring the possibilities for living well that might be opened up by beholding the world at the intersection of human and more than human perspectives, regardless of the ontological register beheld.

The Goodness, Truth, and Beauty of Seeing-with

Syn-theōria, then, provides us with an ideal that takes beholding with animals as a practice through which a worthwhile life might emerge; and by a worthwhile life, I mean one directed toward the good, the true, and the beautiful.[7] Seeing-with is *good* in the sense that it justifies our lives—which is to say, this practice gives normative direction and coherence to the daily and longer-term patterns of living and practicing an ethological disposition. In practicing seeing-with, we strive to become the kinds of beings who are primed to see the world from novel perspectives and with fresh eyes, who spend significant time seeing through the eyes of other creatures, and who delight in such experiences. Such practices serve to displace our tendency to organize the world in all too human terms and toward all too human ends and instead encourage us to hold open possibilities and potentials that arise in encounters with more than human and inhuman modes of existence. To live in this way is to live ethically, in the sense of being vigilant about the form and nature of our subjective, social, and environmental relations, and to consider how we might enhance and enrich those relations rather than degrade them.

The practice of seeing-with also involves a reconfigured conception of *truth*, where truth is understood first and foremost in terms of fidelity and remaining true to ethological encounters and experiments. In this approach, seeing-with is conceived as a way of remaining true to an *event* in Alain Badiou's sense of the term—that is, as a rupture in normal modes of existence that has the effect of calling us to another way of life.[8] From this perspective, deep ethological life can be seen as a response to the event of animals, to the ways animals exceed our conceptual frames and expectations and thereby call us to ways of thinking and living that challenge traditional modes of thought and life. At the same time, this (somewhat unorthodox) notion of truth as fidelity is not entirely divorced from traditional epistemological considerations. Seeing-with-animals and the three ethologies should, to be sure, be informed by the best scientific and traditional accounts of the nature and reality of animal existence and human-animal relations. But in ethological practice of the sort I have outlined here, the goal is not to arrive at a single account of the truth of animals or the world but to explore a multitude of human and more than human perspectives that emerge in response to engagements with an exceedingly rich world. The multiplication of perspectives undertaken by deep ethologists can be seen, then, as one way of gaining a fuller and more accurate picture of reality. Ultimately, though, truth as it is understood and practiced here is cherished most as a way of seeing reality afresh in order to open up other ways of thinking and living well. In this sense, the true and the good interlace toward the end of living worthwhile lives in view of and alongside the kin who render such ways of life possible.

Finally, deep ethological practice and seeing-with animal others give rise to an alternative notion of *the beautiful*. On this theme, we might consider returning to Aristotle—or rather, to the "other," non-anthropocentric Aristotle—who is keen to encourage his readers to study animals in all their richness and variety. Aware that many philosophers might think studying the details of finite animal life less worthy of their time than contemplating the heavenly and divine, Aristotle insists that the animal world is filled with astonishing things that deserve our sustained attention, and that "in all natural things, there is something wonderful."[9] In commenting on this sensibility in Aristotle's work, Pierre Hadot suggests that Aristotle initiates something of an aesthetic revolution in his zoological texts. Rather than limiting wonder and beauty to the kind of order associated with the rationality, ideal function, and proportion of the eternal realm, Aristotle's research on animals suggests (perhaps unwittingly) that the beautiful is to be found even in the mundane details of finite, earthly, animal existence.[10] This worldly "realistic aesthetics" seeks to return our philosophical and reflective gazes back to the earth and to join Heraclitus in insisting that "there are gods here too," in and among the ordinary lives and daily activities of our animal kin.[11]

If this "other" Aristotelian disposition encourages us to turn our attention back to the beauties and wonders of the animal world, many of the contemporary ethological thinkers and practitioners we have been considering ask us to take one step further along this path and to consider the beauty of the world from the perspective of animals themselves. Here the beautiful takes on a novel, inhuman form. We catch sight of the world in the way Linda Hogan (see chapter 3) encourages us to perceive it: as a world not made or intended for us, a world where we are—contra Protagoras—no longer the measure of value or meaning. This is the world as glimpsed from an inhuman perspective, utterly lacking in divine or human telos but not, for all that, any less remarkable, meaningful, or valuable. Seeing-with our animal kin not only grants us partial access to this world but also allows us a glimpse of what it might mean to cocreate inhuman beauty—for animals not only are beholders of this inhuman beauty, but they are agents of and participants in it as well. The often-debated question of whether animals are artists is perhaps beside the point in this context, for there can be little question that animals create and contribute to this inhuman form of beauty in countless ways. The aesthetic question that faces us today is whether we are willing to become worthy of bearing witness to this beauty, and perhaps—if we wish to rise even further to the challenge—to contribute to and increase it rather than further degrade it.

In closing, I must acknowledge that the ideal and practice of *syn-theoria* cannot promise the stable and persistent happiness and flourishing (*eudaimonia*)—the goal of nearly all ancient philosophies—typically associated with *theōria*. Seeing-with animals is, to be sure, the source of some of the most precious joys one can experience, as those who consistently practice deep ethological seeing and relations will eagerly testify. Few things are more rewarding and wonder-filled than coming to appreciate the world from the perspective of another creature. There can be no guarantee, though, that what we see when we inhabit such viewpoints will consistently give rise to such joy. The world is not constituted to ensure the happiness and flourishing of either animals or human beings—a point that holds true in any age, but especially in ours, characterized as it is by widespread and rapid degradation of living conditions across several registers. Of course, even under declining conditions, the earth still bears life richly and manifests astonishing beauty. Yet we cannot deny that, in an age of ecological catastrophe, mass killing, and mass extinction, learning to see the world with and alongside animals will inevitably expose us to the profound grief that attends such destruction.[12] To live in this manner, exposed at the same time to the very worst and the very best, is to live dangerously, to risk everything—but this is a "fine risk to be run," perhaps the finest risk we can take.[13] It is the risk of opening ourselves to our animal and earthly kin in love, resolving to live "wild" lives with them, to remain in community with them, even in the midst of irreversible planetary decline and unthinkable amounts of useless suffering.[14] For those of us desiring to pursue a worthwhile life (and death) with animals in the age we find ourselves in, few other paths remain open.

Acknowledgments

I am grateful to Animal Lives series editors Jane Desmond, Barbara King, and Kim Marra for their thoughtful feedback on and support of this book from conception to completion. I also thank Kyle Wagner for his steadfast efforts in shepherding the manuscript through the review process and for offering encouraging words along the way. Two anonymous reviewers for the University of Chicago Press provided insightful feedback that helped improve the manuscript in important ways. Finally, I dedicate this book to Luce and Christina, daily partners in ethological living.

Notes

Introduction

1. I write "generally" so as not to give the impression of a triumphal ascent of animal studies either inside or outside the academy. The field is not universally viewed as a valid form of discourse or activism, nor has it been immune from being displaced by other critical theoretical frameworks in the "race for theory." That noted, it is undeniable that the field has grown immensely in both presence and significance in a variety of contexts over the past two or more decades.
2. Margo DeMello offers a helpful survey of these trends in *Animals and Society: An Introduction to Human-Animal Studies*, 2nd ed. (New York: Columbia University Press, 2021).
3. Some of the more influential theorists who adopt this affirmative approach are Donna Haraway, Cary Wolfe, Lori Gruen, and Sue Donaldson and Will Kymlicka. In chapter 2, I take up the work of Haraway and of Donaldson and Kymlicka, in particular, at more length.
4. Félix Guattari, *The Three Ecologies*, trans. Ian Pindar and Paul Sutton (New Brunswick, NJ: Athlone Press, 2000).
5. Guattari uses "transversal" in different ways throughout his career. Although I have been influenced by his usage, I primarily emphasize its most basic meaning: that of two things crossing or interacting and thereby forming a new assemblage. In particular I am interested here in allowing the registers of subjectivity, sociality, and environmentality—which are often held apart as distinct domains in both theory and practice—to interact with and mutually inform their respective developments. For more on Guattari's concept of transversality, see Gary Genosko, *Félix Guattari: An Aberrant Introduction* (New York: Continuum, 2002).
6. For a fuller discussion of the relations and differences between *ēthos* (spelled with an eta) and the correlative term *ethos* (spelled with an epsilon) in the ancient Greek context, see Thomas E. Corts, "The Derivation of Ethos," *Speech Monographs* 35 (1968): 201–2; and Charles Chamberlain, "From 'Haunts' to 'Character': The Meaning of *Ēthos* and Its Relation to Ethics," *Helios* 11 (1984): 97–108.

7. See Richard W. Burkhardt, *Patterns of Behavior: Konrad Lorenz, Niko Tinbergen, and the Founding of Ethology* (Chicago: University of Chicago Press, 2005) for a historical account of these founding figures.
8. Nikolaas Tinbergen, "On the Aims and Methods of Ethology," *Zeitschrift für Tierpsychology* 20 (1963): 410–33.
9. Gordon M. Burghardt, "Amending Tinbergen: A Fifth Aim for Ethology," in *Anthropomorphism, Anecdotes, and Animals*, ed. Robert W. Mitchell, Nicholas S. Thompson, and H. Lyn Miles (Albany, NY: SUNY Press, 1997), 254–76.
10. Marc Bekoff, "Deep Ethology, Animal Rights, and the Great Ape/Animal Project: Resisting Speciesism and Expanding the Community of Equals," *Journal of Agricultural and Environmental Ethics* 10 (1998): 269–96.
11. See Becca Franks, Christine Webb, Monica Gagliano, and Barbara Smuts, "Conventional Science Will Not Do Justice to Nonhuman Interests: A Fresh Approach Is Required," *Animal Sentience* 27 (2020): 1–5, for a recent and insightful extension of the argument that animal science needs to develop a more ethical and emotive approach to the study of animals.
12. For a helpful overview of the social and scientific importance of the work of Goodall, Fossey, and Galdikas, see Sy Montgomery, *Walking with the Great Apes: Jane Goodall, Dian Fossey, Biruté Galdikas* (Boston: Houghton Mifflin, 1991); Montgomery's epilogue (256–75) also examines important connections between ethological research and attitudes toward animals in certain Indigenous cultures that anticipate themes I take up at greater length in chapter 3. For more on Barbara Smuts's research, see her *Sex and Friendship in Baboons* (New York: Aldine, 1985). Cynthia Moss provides an overview of her remarkable fieldwork in *Elephant Memories: Thirteen Years in the Life of an Elephant Family* (Chicago: University of Chicago Press, 2000). Caitlin O'Connell chronicles her study of elephant hearing in *The Elephant's Secret Sense: The Hidden Life of the Wild Herds of Africa* (Chicago: University of Chicago Press, 2007); see also her recent work on the overlapping ritual practices of human beings and animals in *Wild Rituals: 10 Lessons Animals Can Teach Us about Connection, Community, and Ourselves* (San Francisco: Chronicle Prism, 2021). I offer my own reflections on the significance of Goodall's and Smuts's work in relation to the philosophical theme of wonder in *Boundaries of Human Nature: The Philosophical Animal from Plato to Haraway* (New York: Columbia University Press, 2022).
13. It would be impossible to provide a comprehensive list of such individuals. For readers seeking an overview of some of these figures and a general introduction to this broader field of research and activism, I recommend any of the following: Jonathan Balcombe, *Second Nature: The Inner Lives of Animals* (New York: Palgrave Macmillan, 2010); Marc Bekoff, *The Emotional Lives of Animals: A Leading Scientist Explores Animal Joy, Sorrow, and Empathy—and Why They Matter* (Novato, CA: New World Library, 2007); Richard Louv, *Our Wild Calling: How Connecting with Animals Can Transform Our Lives—and Save Theirs* (Chapel Hill, NC: Algonquin Books of Chapel Hill, 2019); Carl Safina, *Beyond Words: What Animals Think and Feel* (New York: Henry Holt, 2015).
14. For an overview of this research, see Roberto Marchesini and Marco Celentano, *Critical Ethology and Post-anthropocentric Ethics: Beyond the Separation between Humanities and Life Sciences* (Cham, Switzerland: Springer, 2021).
15. A representative selection of Lestel's work in English translation can be found in Matthew Chrulew, Jeffrey Bussolini, and Brett Buchanan, eds., *The Philosophical Ethology of Dominique Lestel* (New York: Routledge, 2018).

16. Dominique Lestel, Florence Brunois, and Florence Gaunet, "Etho-ethnology and Ethno-ethology," *Social Science Information* 45 (2006): 155–77.
17. Vinciane Despret, "The Body We Care For: Figures of Anthropo-Zoo-Genesis," *Body & Society* 10 (2004): 111–34.
18. See Vinciane Despret, *Bêtes et hommes* (Paris: Gallimard, 2007). A representative selection of Despret's work in English translation can be found in Brett Buchanan, Matthew Chrulew, and Jeffrey Bussolini, eds., *The Philosophical Ethology of Vinciane Despret* (New York: Routledge, 2018).
19. I should also make note of the related and recent trend of *field philosophy*, which seeks to resituate philosophical inquiry in relation to various kinds of "fieldwork" in the human and more than human worlds. At issue for philosophers who adopt this approach is trying to link philosophy more closely to real-world problems and lived experience. For an exploration of this philosophical approach in relation to animal studies and the environmental humanities in particular, see Brett Buchanan, Michelle Bastian, and Matthew Chrulew, eds., "Field Philosophy and Other Experiments," *parallax* 24, no. 4 (2018). Finally, the influence of Gilles Deleuze on the development of philosophical ethology should not be understated. For extensive discussion of the ethological dimensions of Deleuze's philosophy, see Jason Cullen, *Deleuze and Ethology: A Philosophy of Entangled Life* (New York: Bloomsbury Academic, 2021); and Felice Cimatti, *Unbecoming Human: Philosophy of Animality After Deleuze* (Edinburgh: Edinburgh University Press, 2020). I briefly discuss the work of Deleuze (and his coauthor Félix Guattari) in chapter 1.
20. The term ontological refers to the basic, most fundamental mode of existence characteristic of a given entity or group of entities. In this book, my primary concern will be with the ontological relationship between human beings and animals and how to think about similarities and differences between and among these groups.
21. I discuss this concept more fully in Matthew Calarco, *Thinking through Animals: Identity, Difference, Indistinction* (Stanford, CA: Stanford University Press, 2015).

Chapter One

1. Joe Hutto, *Touching the Wild: Living with the Mule Deer of Deadman Gulch* (New York: Skyhorse, 2014), vii; citations from this book will henceforth be given in parentheses in the text. A PBS Nature documentary was made of Hutto's project: Joe Hutto, *Touching the Wild: Living with the Mule Deer of Deadman Gulch*, DVD, dir. David Allen (Alexandria, VA: PBS Home Video, 2014). I urge readers to watch the documentary, which offers a superb overview of Hutto's project and includes some of the most stunning human-deer encounters ever captured on film.
2. Joe Hutto, *The Light in High Places: A Naturalist Looks at Wyoming Wilderness—Rocky Mountain Bighorn Sheep, Cowboys, and Other Rare Species* (New York: Skyhorse, 2009).
3. Hutto, *Touching the Wild*, DVD.
4. For fuller reflections on the issue of animal awareness of death, see Susana Monsó and Antonio J. Osuna-Mascaró, "Death Is Common, So Is Understanding It: The Concept of Death in Other Species," *Synthese* 199 (2021): 2251–75.
5. Although I am using the term deterritorialization in view of Deleuze and Guattari's work, I do not mean to suggest that my use of the term corresponds precisely with theirs. See Gilles Deleuze and Félix Guattari, *A Thousand Plateaus: Capitalism and Schizophrenia*, trans. Brian Massumi (Minneapolis: University of Minnesota Press,

1987), and their *What Is Philosophy?*, trans. Hugh Tomlinson and Graham Burchell (New York: Columbia University Press, 1994).

6. For more on Deleuze and Guattari's notion of becoming-animal, see their *A Thousand Plateaus* and *What Is Philosophy?*, as well as their *Kafka: Toward a Minor Literature*, trans. Dana Polan (Minneapolis: University of Minnesota Press, 1986). I examine this concept at more length in "Being-toward-Meat: An Analytic of Human-Animal Finitude," in *The Animal Inside: Essays at the Intersection of Philosophical Anthropology and Animal Studies*, ed. Geoffrey Dierckxsens, Michael Begun, Rudmer Bijlsma, and Thomas Kiefer (Lanham, MD: Rowman and Littlefield International, 2016), 197–212.
7. Deleuze and Guattari rigorously distinguish becoming-animal from the act of imitating animals (*A Thousand Plateaus*, 305). Hutto's experimental approach blurs this line to some extent.
8. For more on the concept of speciesism, see Richard Ryder, *Animal Revolution: Changing Attitudes towards Speciesism* (Oxford: Basil Blackwell, 1989), and Peter Singer, *Animal Liberation* (New York: Ecco, 2002).
9. I explore the limitations of the standard analysis of speciesism in more depth in "Reorienting Strategies for Animal Justice," in *Philosophy and the Politics of Animal Liberation*, ed. Paola Cavalieri (New York: Palgrave Macmillan, 2016), 45–70.
10. For historical surveys of anthropocentrism, see Robert Renehan, "The Greek Anthropocentric View of Man," *Harvard Studies in Classical Philology* 85 (1981): 239–59; and John Rodman, "Paradigm Change in Political Science: An Ecological Perspective," *American Behavioral Scientist* 24 (1980): 49–78.
11. For more on this point, see Renehan, "Greek Anthropocentric View of Man."
12. This latter point is developed forcefully by Zakiyyah Iman Jackson in *Becoming Human: Matter and Meaning in an Antiblack World* (New York: New York University Press, 2020). In this work, Jackson pushes back against the idea postulated by some animal studies theorists that "human-animal binarism is the original and foundational paradigm upon which discourses of human difference, including, or even especially, racialization was erected" (12). I believe she is right to do so. Anthropocentrism, as articulated here, functions by simultaneously distinguishing the human from a variety of "others," including animals, other human beings, nature, artifacts, and so on. It is crucial to bear in mind these varied divisions that constitute the human, not just in order to understand the history and present of anthropocentric oppression but also in view of the task of developing a genuinely post-anthropocentric way of life that does not repeat the sorts of fractures internal to the human to which Jackson recalls us.
13. Key writings from this period are excerpted in Michel Foucault, *The Foucault Reader*, ed. Paul Rabinow (New York: Pantheon Books, 1984). See also Michel Foucault, *Power/Knowledge: Selected Interviews and Other Writings, 1972–1977*, ed. Colin Gordon (New York: Pantheon Books, 1980).
14. In addition to Foucault's *The History of Sexuality*, vol. 3, *The Care of the Self*, trans. Robert Hurley (New York: Vintage, 1988), see his *The Courage of Truth: The Government of Self and Others II, Lectures at the Collège de France, 1983–1984*, trans. Graham Burchell (New York: Palgrave Macmillan, 2011); and Foucault, *The Hermeneutics of the Subject: Lectures at the Collège de France 1981–1982*, trans. Graham Burchell (New York: Palgrave Macmillan, 2005).
15. Michel Foucault, *The History of Sexuality*, vol. 2, *The Use of Pleasure*, trans. Robert Hurley (New York: Random House, 1985), 8.

16. For more on Hadot's understanding of philosophy as a series of spiritual exercises, see his *Philosophy as a Way of Life: Spiritual Exercises from Socrates to Foucault*, ed. Arnold Davidson (Oxford: Wiley-Blackwell, 1995).
17. Foucault, *Courage of Truth*.
18. Foucault, 264.
19. Foucault.
20. These are common themes among ancient Cynics. See William Desmond's *Cynics* (Berkeley: University of California Press, 2008) for an expert survey of Cynic philosophy as well as several helpful comments about Cynic attitudes toward animals and the more than human world.
21. I present a detailed analysis of this minor tradition in a forthcoming work on veganism as a spiritual exercise.

Chapter Two

1. Although gift-giving by crows is not widely reported either anecdotally or in the scientific literature, it is documented frequently enough to be taken seriously by crow ethologists. John Marzluff and Tony Angell discuss the issue in *Gifts of the Crow: How Perception, Emotion, and Thought Allow Smart Birds to Behave Like Humans* (New York: Free Press, 2012), 108–15.
2. Nathan J. Emery and Nicola S. Clayton, "The Mentality of Crows: Convergent Evolution of Intelligence in Corvids and Apes," *Science* 306 (2004): 1903–7.
3. For an extensive and careful study of tool use by crows, see Lucas A. Bluff et al., "Tool Use by Wild New Caledonian Crows [*Corvus moneduloides*] at Natural Foraging Sites," *Proceedings of the Royal Society B: Biological Sciences* 277 (2010): 1377–85.
4. Auguste Marie Philippa von Bayern et al., "Compound Tool Construction by New Caledonian Crows," *Scientific Reports* 8 (2018): 1–8.
5. The classic study on this topic is Dwight R. Chamberlain and George W. Cornwell, "Selected Vocalizations of the Common Crow," *Auk* 88 (1971): 613–34.
6. Kaeli N. Swift and John M. Marzluff, "Wild American Crows Gather around Their Dead to Learn about Danger," *Animal Behaviour* 109 (2015): 187–97.
7. Lawrence Kilham, "Cooperative Breeding of American Crows," *Journal of Field Ornithology* 55 (1984): 349–56.
8. Lisa Horn et al., "Sex-Specific Effects of Cooperative Breeding and Colonial Nesting on Prosociality in Corvids," *eLife* 9 (2020): e58139.
9. I don't want to give the impression that there is no conflict or hostility among the crows in our neighborhood or crows more generally. There is plenty of fighting among crows, especially during mating season and in spring when fledglings are in the nest. Note also that crows are usually adept at dealing with such conflict and finding ways to reconcile. For a study of conflict resolution in carrion crows, see Miriam Jennifer Sima et al., "Reconciliation and Third-Party Affiliation in Carrion Crows," *Ethology* 124 (2018): 33–44.
10. For a helpful introduction to these ongoing debates, see Kevin N. Laland and Bennett G. Galef, eds., *The Question of Animal Culture* (Cambridge, MA: Harvard University Press, 2009). Marine biologists Hal Whitehead and Luke Rendell offer a lively survey of the animal culture "wars" from a perspective consonant with the one I develop in this chapter; see *The Cultural Lives of Whales and Dolphins* (Chicago: University of Chicago Press, 2015).

11. See John Marzluff and Tony Angell, "Cultural Coevolution: How the Human Bond with Crows and Ravens Extends Theory and Raises New Questions," *Journal of Ecological Anthropology* 9 (2005): 69–75. This argument is expanded in John Marzluff and Tony Angell, *In the Company of Crows and Ravens* (New Haven, CT: Yale University Press, 2005).
12. For provocative experiments with and reflections on such relations, see Jim Dutcher and Jamie Dutcher, *The Wisdom of Wolves: Lessons from the Sawtooth Pack* (Washington, DC: National Geographic, 2018); and Timothy Treadwell and Jewel Palovak, *Among Grizzlies: Living with Wild Bears in Alaska* (New York: Ballantine, 1997). I thank Blake Ginsburg for helpful conversations on this issue over the past several years.
13. I use the phrase *gay science* (*fröhliche Wissenschaft*) in the sense that Friedrich Nietzsche lends it in *The Gay Science: With a Prelude in Rhymes and an Appendix of Songs*, trans. Walter Kaufmann (New York: Vintage Books, 1974). For Nietzsche, "gay" or "joyful" science is a mode of knowing that seeks to engage and experiment with the world toward the end of joint flourishing (as opposed to gaining knowledge for its own sake or to make existence easier and more comfortable).
14. See Guattari, *Three Ecologies*, 47–52.
15. For more on the concept of the "animal-industrial complex," see Barbara Noske, *Humans and Other Animals: Beyond the Boundaries of Anthropology* (London: Pluto, 1989), and Richard Twine, "Revealing the 'Animal-Industrial Complex': A Concept and Method for Critical Animal Studies?," *Journal for Critical Animal Studies* 10 (2012): 12–39. On the question of animal labor, see Kendra Coulter's expert overview in *Animals, Work, and the Promise of Interspecies Solidarity* (New York: Palgrave Macmillan, 2016).
16. Hesiod, *Works and Days*, trans. Catherine M. Schlegel and Henry Weinfield (Ann Arbor: University of Michigan Press, 2006), 277–79.
17. For a critical overview of this anthropocentric tradition, see Gary Steiner, *Anthropocentrism and Its Discontents: The Moral Status of Animals in the History of Western Philosophy* (Pittsburgh, PA: University of Pittsburgh Press, 2005).
18. This list is not meant to be exhaustive of all pro-animal philosophers; rather, it names important figures for building a kind of counter-history of animal philosophy. I discuss a number of these figures in a forthcoming volume titled *Kinship: A Counter-History of Animal Philosophy*.
19. Although in the first two chapters of this book I rely primarily on (minor) Western philosophical and theoretical traditions and thinkers to develop a critical ethological framework, in the following chapter I show how this counter-historical current opens beyond itself to ongoing conversations in other traditions, especially Native American Indigenous traditions. At stake in my approach is an effort to constitute the conditions of possibility for these minor Western traditions to enter into a genuine dialogue and a critical, transversal relation with other traditions that have often been overlooked, disavowed, or even deliberately and violently suppressed by the established order.
20. Sue Donaldson and Will Kymlicka, *Zoopolis: A Political Theory of Animal Rights* (New York: Oxford University Press, 2011).
21. I'm waiting until chapter 3 to discuss the problem of settler colonialism in the context of liberalism. For a critique of the lingering presence of settler colonialism in Donaldson and Kymlicka's work, see Billy-Ray Belcourt, "Animal Bodies, Colonial Subjects: (Re)locating Animality in Decolonial Thought," *Societies* 5 (2014): 1–11.

For a related, in-depth critical analysis of Donaldson and Kymlicka's project, see Dinesh J. Wadiwel, *The War against Animals* (Boston: Brill, 2015).

22. This section contains elements from chapter 13 of my book *The Boundaries of Human Nature: The Philosophical Animal from Plato to Haraway* (New York: Columbia University Press, 2021).
23. See Donna J. Haraway, *Companion Species Manifesto* (Chicago: Prickly Paradigm Press, 2003); and Haraway, *When Species Meet* (Minneapolis: Minnesota University Press, 2008). Haraway uses the term companion species in an expansive sense to include human-animal relations as well as more than human relations of many kinds. In her recent work, she often uses the terms multispecies relations and compost to explore cross-species engagements and to stress the roles of both generation and decay in such relations (see, for example, Donna Haraway, *Staying with the Trouble: Making Kin in the Chthulucene* [Minneapolis: University of Minnesota Press, 2016]). I am focusing on the term companion species here, since it is more commonly used in discussions of Haraway's work by animal studies theorists and is, I hope, already familiar to some readers.
24. For Haraway, nature and culture are co-constitutive and overlapping registers of reality rather than oppositional and homogenous spheres.
25. I borrow this use of the term claustrophobic from Margaret Grebowicz's insightful essay, "Orca Intimacies and Environmental Slow Death: Earthling Ethics for a Claustrophobic World," in *Routledge Handbook of Gender and Environment*, ed. Sherilyn MacGregor (London: Routledge, 2017), 115–30.
26. Haraway, *Companion Species Manifesto*, 46.
27. See, for example, Carol J. Adams, "An Animal Manifesto: Gender, Identity, and Vegan-Feminism in the Twenty-First Century," *Parallax* 12 (2006): 120–28; Paula Arcari, Fiona Probyn-Rapsey, and Haley Singer, "Where Species Don't Meet: Invisibilized Animals, Urban Nature, and City Limits," *Environment and Planning E: Nature and Space* 4 (2021): 940–65; and Zipporah Weisberg, "The Broken Promises of Monsters: Haraway, Animals, and the Humanist Legacy," *Journal for Critical Animal Studies* 7 (2009): 21–61.
28. Haraway makes these points in several places, but most clearly and succinctly in her interview with Jeffrey J. Williams, "Science Stories: An Interview with Donna J. Haraway," *Minnesota Review* 73–74 (2010): 133–63. When pressed on her ideas about veganism, Haraway has refined her views to some extent. See, for example, her remarks on veganism and pastoralism in Sarah Franklin, "Staying with the Manifesto: An Interview with Donna Haraway," *Theory, Culture and Society* 34 (2017): 49–63, 56.
29. Readers interested in exploring these issues at more length should consult Philippe Descola, *Beyond Nature and Culture*, trans. Janet Lloyd (Chicago: University of Chicago Press, 2013); and Samantha Hurn, *Humans and Other Animals: Cross-Cultural Perspectives on Human-Animal Interactions* (London: Pluto Press, 2012).
30. Jennifer Wolch, "Zoöpolis," in *Animal Geographies: Places, Politics, and Identity in the Nature-Culture Borderlands*, ed. Jennifer Wolch and Jody Emel (London: Verso, 1998), 119–38.
31. See Nigel Thrift, *Killer Cities* (Thousand Oaks, CA: Sage, 2021).
32. Wolch, "Zoöpolis," 124.
33. Wolch, 124.
34. Donaldson and Kymlicka underscore their desire to differentiate their approach

from "the more enthusiastic descriptions some authors give of the place of liminal animals." They resist Wolch's suggestion "that humans have a duty to 'invite animals back in,'" suggesting that "we can take reasonable steps to keep would-be opportunistic animals out"; and they further insist that the goal of their work is not "to create a more 'cohesive' community with them" (*Zoopolis*, 299, n. 46). These remarks about liminal animals clearly imply that Donaldson and Kymlicka would be even less enthusiastic about the notion of inviting "wild" animals into what they consider properly human spaces.

35. Eva H. Giraud, *What Comes after Entanglement? Activism, Anthropocentrism, and an Ethics of Exclusion* (Durham, NC: Duke University Press, 2019).

Chapter Three

1. Linda Hogan, *The Radiant Lives of Animals* (Boston: Beacon Press, 2020); I will cite this work parenthetically in the text. For readers new to Hogan's work, I highly recommend her earlier essay "First People," in *Intimate Nature: The Bond between Women and Animals*, ed. Linda Hogan, Deena Metzger, and Brenda Peterson (New York: Fawcett, 1998), 6–19, which encapsulates many of the themes discussed in this chapter.
2. In this passage, Hogan is referring specifically to a scene in one of her novels, *Power* (New York: W. W. Norton, 1998), 191. I should emphasize that, for Hogan, living in this "beautiful and whole" manner is no utopian ideal but is intended to recall, in line with the narratives of several Indigenous traditions, earlier ages when human beings had better and closer relations with animals. Fruitful parallels can be found in classical Western authors (for example, Hesiod, Plato, Virgil, Ovid, and others) who refer to a previous golden age of human existence when human beings lived in remarkable ways that were more aligned with animals and nature.
3. This theme of beholding is examined at more length in the conclusion, in relation to the philosophical concept of *theōria*.
4. Aldo Leopold, *A Sand County Almanac, and Sketches Here and There* (Oxford: Oxford University Press, 1989), 204.
5. An anonymous reviewer of this book in its early stages suggested that, in addition to an analysis of extinction, this section should also include a fuller discussion of defaunation (which is typically understood to refer to the loss not just of animal species but of populations as well as the reduction of the number of individuals of a given species). For strategic reasons, I limit my focus here to extinction, although I agree with the reviewer that defaunation (and its related crises of land and ocean degradation and the overall decline of wildlife) is also of great importance for understanding the stakes of developing an environmental ethology. For a scientific overview of defaunation and related ecological issues, see Rodolfo Dirzo et al., "Defaunation in the Anthropocene," *Science* 345 (2014): 401–6. For a discussion of these issues from a perspective consonant with the argument developed here, see Eileen Crist, "Animals," in *The Cambridge Companion to Literature and the Anthropocene*, ed. John Parham (New York: Cambridge University Press, 2021), 196–210, as well as her more fully developed analysis of these issues in *Abundant Earth: Toward an Ecological Civilization* (Chicago: University of Chicago Press, 2019).
6. Jurriaan M. De Vos, Lucas N. Joppa, John L. Gittleman, and Patrick R. Stephens, "Estimating the Normal Background Rate of Species Extinction," *Conservation Biology* 29 (2015): 452–62.
7. For discussion of a range of estimates, see Gerardo Ceballos et al., "Accelerated

Modern Human-Induced Species Losses: Entering the Sixth Mass Extinction," *Science Advances* 1 (2015): 1–5.

8. Although I employ the commonly used term anthropogenic, I am not suggesting that the problems at issue here are equally generated by human beings per se. The extinction event, as well as the acceleration of climate change to be discussed below, is generated in large part by wealthy residents of advanced capitalist societies. For further discussion of this point, see Andreas Malm and Alf Hornborg, "The Geology of Mankind? A Critique of the Anthropocene Narrative," *Anthropocene Review* 1 (2014): 62–69.
9. Eileen Crist, "Ecocide and the Extinction of Animal Minds," in *Ignoring Nature No More: The Case for Compassionate Conservation*, ed. Marc Bekoff (Chicago: University of Chicago Press, 2013), 45–61.
10. Vinciane Despret, "'It Is an Entire World That Has Disappeared,'" in *Extinction Studies: Stories of Time, Death and Generations*, ed. Deborah Bird Rose, Thom van Dooren, and Matthew Chrulew (New York: Columbia University Press, 2017), 217–22.
11. Dominique Lestel, "The Withering of Shared Life through the Loss of Biodiversity," *Social Science Information* 52 (2013): 307–25, 320.
12. Thom van Dooren, *Flight Ways: Life and Loss at the Edge of Extinction* (New York: Columbia University Press, 2014); see also his *The Wake of Crows: Living and Dying in Shared Worlds* (New York: Columbia University Press, 2019).
13. For a brilliant example of what such work might look like in view of the ongoing effects of colonialism on both human and more than human beings (concerns that form the focus of the second portion of this chapter), see Juno Salazar Parreñas, *Decolonizing Extinction: The Work of Care in Orangutan Rehabilitation* (Durham, NC: Duke University Press, 2018).
14. Jessica Blunden and Tim Boyer, eds., "State of the Climate in 2020," *Bulletin of the American Meteorological Society* 102 (2020), Si–S475.
15. Intergovernmental Panel on Climate Change, *Climate Change 2021: The Physical Science Basis; Contribution of Working Group I to the Sixth Assessment Report of the Intergovernmental Panel on Climate Change* (New York: Cambridge University Press, 2021), 5.
16. P. J. Gerber, H. Steinfeld, B. Henderson, et al., *Tackling Climate Change through Livestock: A Global Assessment of Emissions and Mitigation Opportunities* (Rome: Food and Agriculture Organization of the United Nations, 2013).
17. For more in-depth research on these points, see Anders Nordgren, "Ethical Issues in Mitigation of Climate Change: The Option of Reduced Meat Production and Consumption," *Journal of Agricultural and Environmental Ethics* 25 (2012): 563–84; Peter Scarborough, Paul N. Appleby, Anja Mizdrak et al., "Dietary Greenhouse Gas Emissions of Meat-Eaters, Fish-Eaters, Vegetarians, and Vegans in the UK," *Climatic Change* 125 (2014): 1–14; Marco Springmann, H. Charles, J. Godfray, Mike Rayner et al., "Analysis and Valuation of the Health and Climate Change Cobenefits of Dietary Change," *Proceedings of the National Academy of Sciences of the United States of America* 113 (2016): 4146–51; Tony Weis, *The Ecological Hoofprint: The Global Burden of Industrial Livestock* (London: Zed, 2013); Juliana Minetto, Gellert Paris et al., "Changing Dietary Patterns Is Necessary to Improve the Sustainability of Western Diets from a One Health Perspective," *Science of the Total Environment* 811 (2022): 1–10.
18. For an extensive overview of these issues, see Intergovernmental Panel on Climate Change, *Climate Change 2014: Impacts, Adaptation, and Vulnerability, Part A* (Cambridge: Cambridge University Press, 2014).

19. See Kim TallBear, "Why Interspecies Thinking Needs Indigenous Standpoints," *Fieldsights*, November 18, 2011, https://culanth.org/fieldsights/why-interspecies-thinking-needs-indigenous-standpoints.
20. This point is also made by Zoe Todd in "An Indigenous Feminist's Take on the Ontological Turn: 'Ontology' Is Just Another Word for Colonialism," *Journal of Historical Sociology* 29 (2016): 4–22.
21. TallBear, "Why Interspecies Thinking Needs Indigenous Standpoints."
22. This theme is examined thoughtfully and at length by Keith Basso in his *Wisdom Sits in Places: Landscape and Language among the Western Apache* (Albuquerque: University of New Mexico Press, 1996).
23. Glen Sean Coulthard, *Red Skin, White Masks: Rejecting the Colonial Politics of Recognition* (Minneapolis: University of Minnesota Press, 2014).
24. Coulthard, 61.
25. Coulthard, 61.
26. Coulthard, 12–13.
27. Coulthard, 13.
28. Coulthard, 13.
29. Ursula K. Heise, *Sense of Place and Sense of Planet: The Environmental Imagination of the Global* (New York: Oxford University Press, 2008), 21.
30. For further reflections on this complex set of issues, see Brian Yazzie Burkhart, *Indigenizing Philosophy through the Land: A Trickster Methodology for Decolonizing Environmental Ethics and Indigenous Futures* (East Lansing: Michigan State University Press, 2019).
31. Robin Wall Kimmerer, "The Rights of the Land," *Orion Magazine*, November–December 2008. https://orionmagazine.org/article/the-rights-of-the-land/. For Kimmerer's fuller discussion of this lawsuit and related themes, see *Braiding Sweetgrass* (Minneapolis: Milkweed Editions, 2013), 310–40. Kimmerer's insightful examination of the concept of land in this book serves as an ideal introduction for readers new to this discussion. The Onondaga's land claim was ultimately rejected by the courts, and the EPA has continued cleanup efforts in the region.
32. Kimmerer, "The Rights of the Land."
33. Kimmerer. For further reflections on this point, see Eva Mackey, *Unsettled Expectations: Uncertainty, Land and Settler Decolonization* (Halifax: Fernwood, 2016); and Denise M. Nadeau, *Unsettling Spirit: A Journey into Decolonization* (Montreal: McGill-Queen's University Press, 2020).
34. Belcourt, "Animal Bodies, Colonial Subjects."
35. Fiona Probyn-Rapsey and Lynette Russell, "Indigenous, Settler, Animal: A Triadic Approach," *Animal Studies Journal* 11 (2022): 38–68.
36. Maneesha Deckha, *Animals as Legal Beings: Contesting Anthropocentric Legal Orders* (Toronto: University of Toronto Press, 2020). See also Juliette Singh, *Unthinking Mastery: Dehumanism and Decolonial Entanglements* (Durham, NC: Duke University Press, 2017).
37. The conference "Decolonizing Critical Animal Studies, Cripping Animal Studies" was held in 2016; the proceedings were published in part in Kelly Struthers Montford and Chloë Taylor, eds., *Colonialism and Animality: Anti-colonial Perspectives in Critical Animal Studies* (New York: Routledge, 2020). The New Zealand Centre for Human-Animal Studies hosted the conference "Decolonizing Animals" in 2019. In 2019 the journal *Humanimalia* published a special issue titled "Decolonizing Human-Animal Studies."

38. For further discussion of activism along these lines, see R.D., "Vegan-Washing Genocide: Animal Advocacy on Stolen Land and Re-imagining Animal Liberation as Anti-colonial Praxis," in *Undoing Human Supremacy: Anarchist Political Ecology in the Face of Anthroparchy*, ed. Simon Springer, Jennifer Mateer, and Martin Locret-Collet (Lanham, MD: Rowman and Littlefield, 2021), 89–118.
39. On this point, see Carol J. Adams and Lori Gruen, "Ecofeminist Footings," in *Ecofeminism: Feminist Intersections with Other Animals and the Earth*, 2nd ed., ed. Carol J. Adams and Lori Gruen (New York: Bloomsbury Academic, 2022), 1–46.
40. See especially Val Plumwood, *Feminism and the Mastery of Nature* (New York: Routledge, 1993); Plumwood, *Environmental Culture: The Ecological Crisis of Reason* (New York: Routledge, 2002); and Plumwood, *The Eye of the Crocodile*, ed. Lorraine Shannon (Canberra: Australian National University e-Press, 2012). I thank an anonymous reviewer for the University of Chicago Press for providing the impetus to discuss the points in this paragraph further and for providing many helpful suggestions to this end, several of which I have included here.
41. This point is thoughtfully explored by Kelsey Dayle John (Diné), "Animals," in *Keywords in Radical Philosophy and Education: Common Concepts for Contemporary Movements*, ed. Derek R. Ford (Boston: Brill Sense, 2019), 7–18.
42. Joe Hutto, *Touching the Wild: Living with the Mule Deer of Deadman Gulch* (New York: Skyhorse, 2014), 77.
43. Margaret Robinson, "Veganism and Mi'kmaq Legends," *Canadian Journal of Native Studies* 33 (2013): 189–96; henceforth cited parenthetically in the text.
44. For more on the history behind the (settler) colonial present in relation to food, see Kristin Burnett, Travis Hay, and Lori Chambers, "Settler Colonialism, Indigenous Peoples, and Food: Federal Indian Policies and Nutrition Programs in the Canadian North since 1945," *Journal of Colonialism and Colonial History* 17 (2016): doi:10.1353/cch.2016.0030.
45. A helpful discussion of this sort of approach to maintaining traditional foodways is found in David S. Walsh, "They Call Us 'Caribou Eaters': Negotiating Tłı̨chǫ Dene Relationships with Caribou," in *Native Foodways: Indigenous North American Religious Traditions and Foods*, ed. Michelene E. Pesantubbee and Michael J. Zogry (Albany: SUNY Press, 2021), 109–36.
46. In an interview regarding her research on these themes, Robinson notes that her work has generated numerous responses from "excited white people" who have perhaps presumed an easy identification of veganism with indigenism ("All My Relations: An Interview with Margaret Robinson," in *Animal Subjects 2.0*, ed. Jodey Castricano and Lauren Corman [Waterloo, ON: Wilfrid Laurier University Press, 2016], 229–48, 240). As I hope I have shown in this chapter, the intersections here are far more complex and should give rise to sustained, critical reflection on the part of pro-animal discourse and activism and their relation to settler colonialism.

Conclusion

1. Pierre Hadot, *Philosophy as a Way of Life: Spiritual Exercises from Socrates to Foucault*, ed. Arnold I. Davidson, trans. Michael Chase (Malden, MA: Blackwell, 1995), 265.
2. For a helpful discussion of these terms in an ancient Greek context, see Ian Rutherford, *State Pilgrims and Sacred Observers in Ancient Greece: A Study of Theōria and Theōroi* (New York: Cambridge University Press, 2013).
3. I will bypass the thorny and long-standing scholarly question of how the *summum*

bonum of *theōria* relates to the other virtues and whether Aristotle's theory of the good life leaves sufficient room for practical and political wisdom. This question is insightfully addressed by Gabriel Richardson Lear in *Happy Lives and the Highest Good: An Essay on Aristotle's "Nicomachean Ethics"* (Princeton, NJ: Princeton University Press, 2004).

4. Aristotle, *Nicomachean Ethics*, trans. Robert C. Bartlett and Susan D. Collins (Chicago: University of Chicago Press, 2011), 1178b 25–29.
5. The notion that *theōria* circulates beyond the human is neither a modern notion nor original with us. The neo-Platonic philosopher Plotinus, for example, famously argued that all things strive for and participate in *theōria* in their own ways.
6. Here I am alluding to both María Lugones, "Playfulness, 'World'-Travelling, and Loving Perception," *Hypatia* 2 (1987): 3–19, and Jakob von Uexküll, *A Foray into the World of Animals and Humans, with "A Theory of Meaning,"* trans. Joseph O'Neil (Minneapolis: University of Minnesota Press, 2010).
7. These categories, while often associated with the "transcendentals" of medieval philosophy, have much older roots in ancient Greek philosophy. In appropriating these terms, I intend to link the philosophical ethological project articulated here with more standard philosophical paradigms while also challenging the anthropocentrism of much (but certainly not all) of the philosophical tradition.
8. For readers unfamiliar with the formidable philosophy of Alain Badiou, a helpful introduction to his concept of the event and other key ideas in his work can be found in Alain Badiou and Fabien Tarby, *Philosophy and the Event*, trans. Louise Burchill (Cambridge: Polity, 2013). Although I am here making use of his concept of the event, Badiou himself would object to extending the concept to encounters with the nonhuman/animal world in the way I do.
9. Aristotle, "Parts of Animals," trans. D. M. Balme, in *A New Aristotle Reader*, ed. J. L. Ackrill (Princeton, NJ: Princeton University Press, 1987), 645a.
10. I write "perhaps unwittingly," as Aristotle himself justifies the study of mundane animal life primarily through its ability to reveal how all things participate in more permanent structures, and thus his aesthetic is in some ways opposed to the one Hadot is sketching here. That said, Aristotle's ethological sensibilities and sustained attention to the details of animal behavior certainly betray more than an indirect interest in the mundane realities of finite animal existence.
11. Aristotle, "Parts of Animals," 645a.
12. On the issue of grief (both human and more than human), see Margo DeMello, ed., *Mourning Animals: Rituals and Practices Surrounding Animal Death* (East Lansing: Michigan State University Press, 2016); Barbara J. King, *How Animals Grieve* (Chicago: University of Chicago Press, 2013); and Thom van Dooren, *Flight Ways: Life and Loss at the Edge of Extinction* (New York: Columbia University Press, 2014).
13. I borrow the phrases "the very best" and "the very worst" from Jacques Derrida, who uses this language in several places in his writings. To "live dangerously" is an allusion to Friedrich Nietzsche's counsel on living well in *The Gay Science*, §283, as well as to the work of Emmanuel Levinas, from whom I also borrow the phrase "fine risk to be run." Levinas describes the ethical relation with the Other as "a dangerous life, a fine risk to be run" (see Emmanuel Levinas, *Otherwise Than Being, or Beyond Essence*, trans. Alphonso Lingis [Norwell, MA: Kluwer, 1991], 120).
14. On the notion of a "wild" life, see Rosemary-Claire Collard, *Animal Traffic: Lively Capital in the Global Exotic Pet Trade* (Durham, NC: Duke University Press, 2020).

For Collard, a "wild life" is "one in which animals engage in their own life-making practices, until they die. There is no guarantee of flourishing, only the conditions of possibility for a degree of creative self-determination and community" (139). In emphasizing remaining with animals in the midst of the difficulties of existence, I am echoing sentiments found in Donna Haraway's *Staying with the Trouble: Making Kin in the Chthulucene* (Minneapolis: University of Minnesota Press, 2016).

Index